Presenting Australia's National Parks

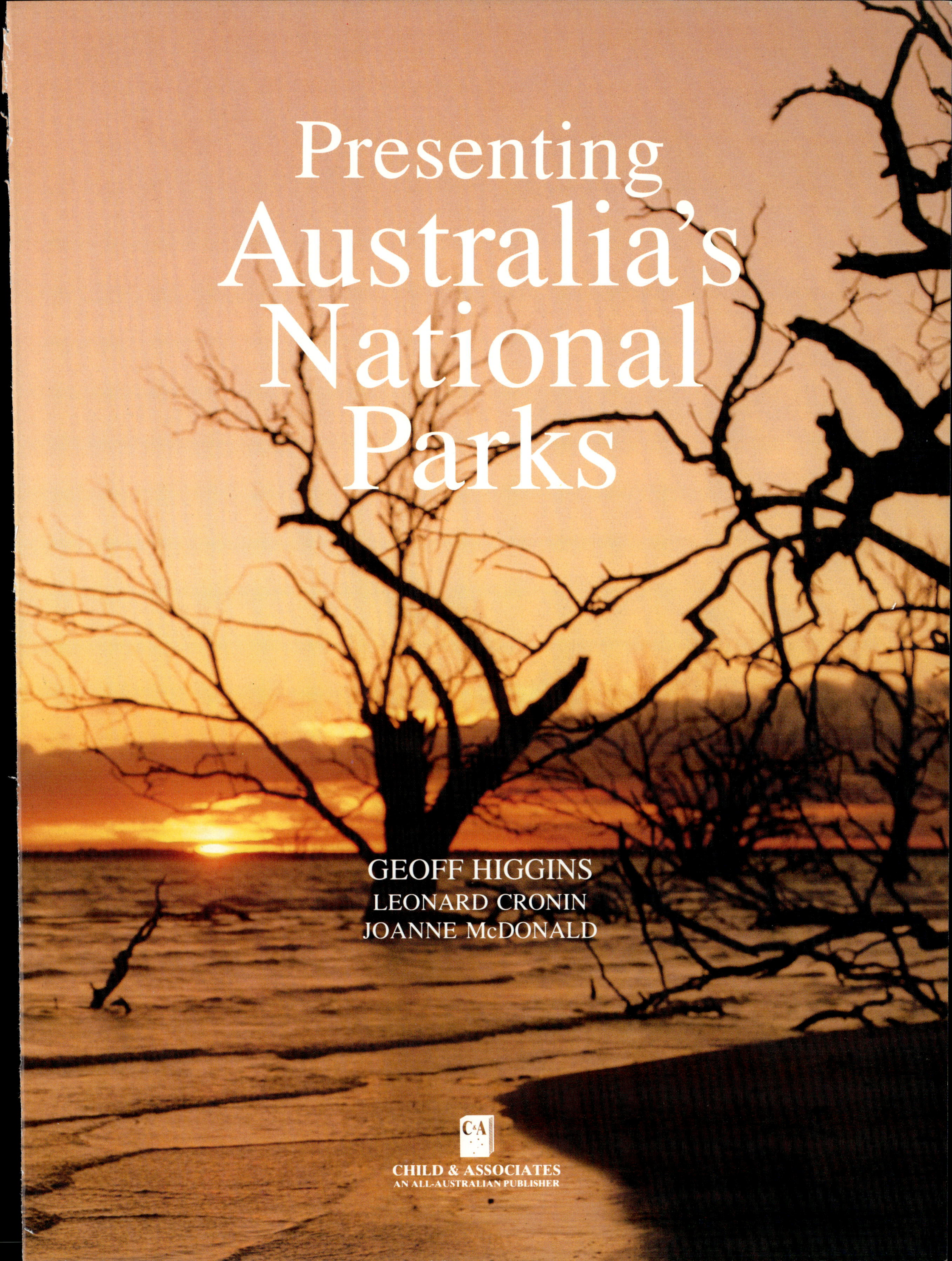

Presenting Australia's National Parks

GEOFF HIGGINS
LEONARD CRONIN
JOANNE McDONALD

CHILD & ASSOCIATES
AN ALL-AUSTRALIAN PUBLISHER

ACKNOWLEDGEMENTS

The publishers wish to express their sincere thanks to the Rangers and Officers of the various National Parks who replied so promptly with updated information and added material.

Particular thanks to Don Marshall of the Queensland National Parks and Wildlife Service.

Half title page: *A friendly greeting after a frolic in the surf at Seal Bay in Flinders Chase National Park.*

Title page: *Sunset over Lake Cawndilla in Kinchega National Park.*

Contents page: *Statuesque Cape du Couedic Lighthouse on the south-western tip of Kangaroo Island in Flinders Chase National Park.*

Published by
Child & Associates Publishing Pty Ltd
5 Skyline Place, Frenchs Forest, NSW, Australia, 2086
A wholly owned Australian publishing company
This book has been edited, designed and typeset in Australia by the Publisher
Typesetting processed by Deblaere Typesetting Pty Ltd
First edition 1986
Second edition, revised and reset, 1988
Photography by Geoff Higgins
Text by Leonard Cronin and Joanne McDonald

Printed in Hong Kong by South Sea International Press Ltd

National Library of Australia
Cataloguing-in-Publication Data

Higgins, Geoff.
Presenting Australia's national parks.

2nd ed.
Includes index.
ISBN 0 86777 269 7.

1. National parks and reserves—Australia—Guide-books. 2. Australia—Description and travel—1976– —Guide-books. I. Cronin, Leonard. II. McDonald, Joanne. III. Title.

919.4'0463

Contents

Introduction

Australia is blessed with a variety of unspoiled landscapes and some of the last true wilderness areas in the world. They vary dramatically from the lush tropical islands of the Barrier Reef, through a magnificent coastline of rocky headlands and endless beaches, to desert lands and alpine forests.

In an attempt to protect these irreplaceable naturelands from encroaching civilisation, large areas, representing some of the most spectacular landscapes of this continent, have been set aside for the enjoyment of all under the management of the state governments. They contain a wealth of wildlife, scenery, and the peace of the great outdoors, and include some 260 National Parks, varying in size from a mere 11 ha to more than 16 000 sq km.

Many are accessible only by off-road vehicles, and remain undeveloped wilderness areas. While these offer so much to the naturalist and dedicated bushwalker, they are often remote, infrequently visited and perhaps best left in this relatively undisturbed state. Others are sufficiently developed to cater for campers, caravanners and picnickers, with well-developed walking and nature trails, lookouts, visitor centres, and a whole range of activities.

To look in detail at all the National Parks in Australia is far beyond the scope of this book. The amount of information is so great that it would fill many volumes and overwhelm all but the most dedicated naturalist. Instead, we have carefully selected those National Parks which are readily accessible by conventional vehicles, offer camping facilities in the park or nearby, picnic areas, walking tracks, and a variety of scenery, wildlife and activities of interest to the average nature lover and family group.

With the help of stunning photography, maps, and a concise, carefully researched text, we have produced a practical, beautifully illustrated guide and indispensable companion for those interested in the great Australian countryside.

Some eighty-two National Parks fall into the above category. These we describe individually, state-by-state, in sufficient detail to enable the reader to clearly identify the landscapes, wildlife, climate, facilities, activities and accessibility of each park. We recommend the best seasons for visitors and whether bookings are necessary. We take a look at the most commonly encountered animals, birds, reptiles, trees and wildflowers.

A large scale map of Australia indicates the location of every featured park. All this is complemented by beautiful photography bringing to life the geography, flora and fauna of our natural wonderlands.

The Visitor's Guide lists the remaining National Parks in alphabetical order, state-by-state, and indicates such facilities as the types of campsites and other accommodation available, visitor centres, walking tracks, water availability and any access problems. This ready-reference guide will prove invaluable to those planning tours or excursions.

It is hoped that by providing this information in a well presented and easily accessible form more people will come to love, appreciate and respect our fragile heritage. By experiencing for ourselves the natural wonders of this great continent we will be able and more willing to understand the concern of environmentalists who are fighting to preserve our countryside from the ever encroaching developers, and those who would plunder our forests and denude the landscape to satisfy their material greed.

Spectacular coastline views can be seen from Torndirrup National Park.

KEY TO MAP

New South Wales

Queensland

Victoria

South Australia

Northern Territory

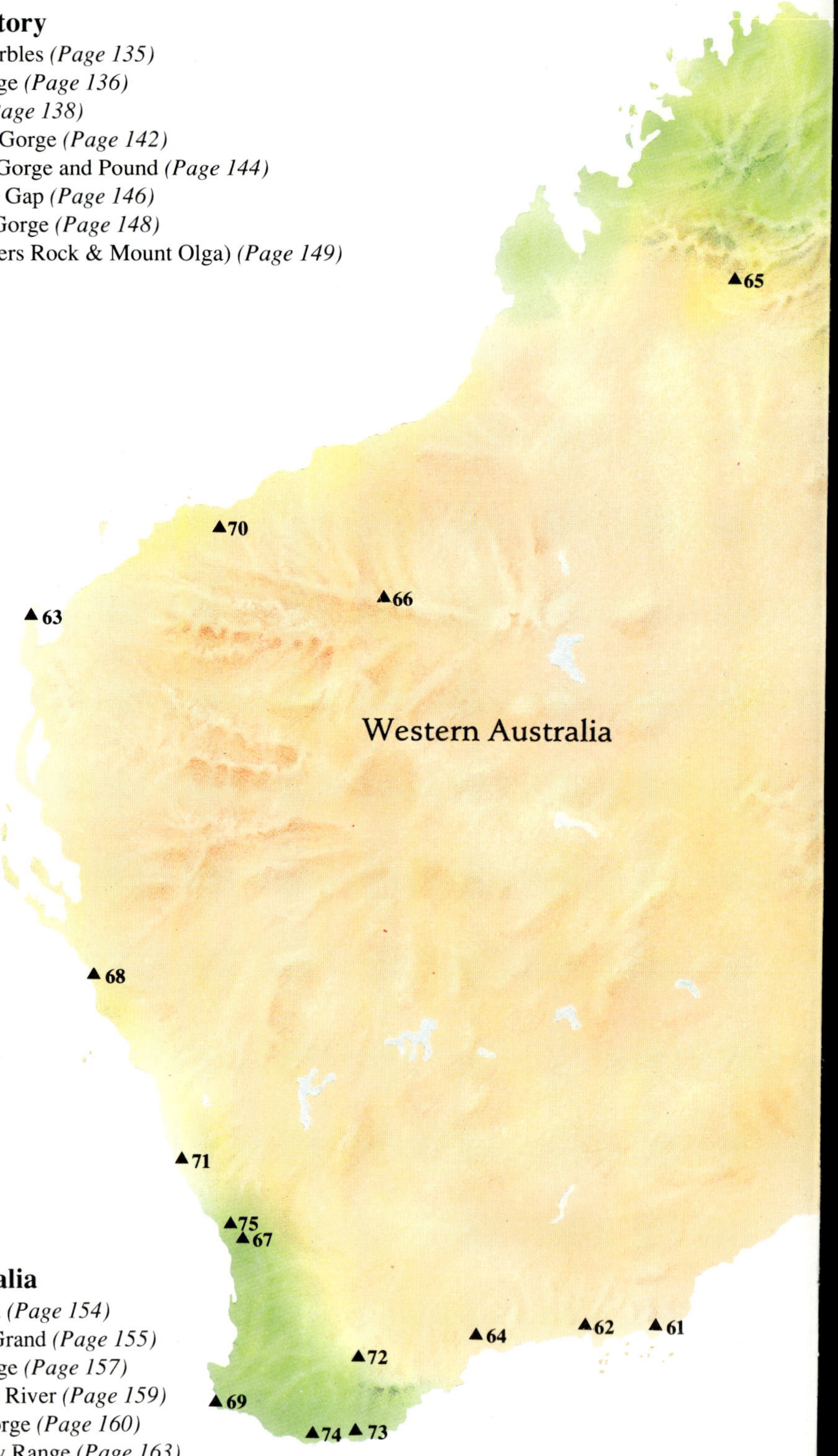

Western Australia

Tasmania

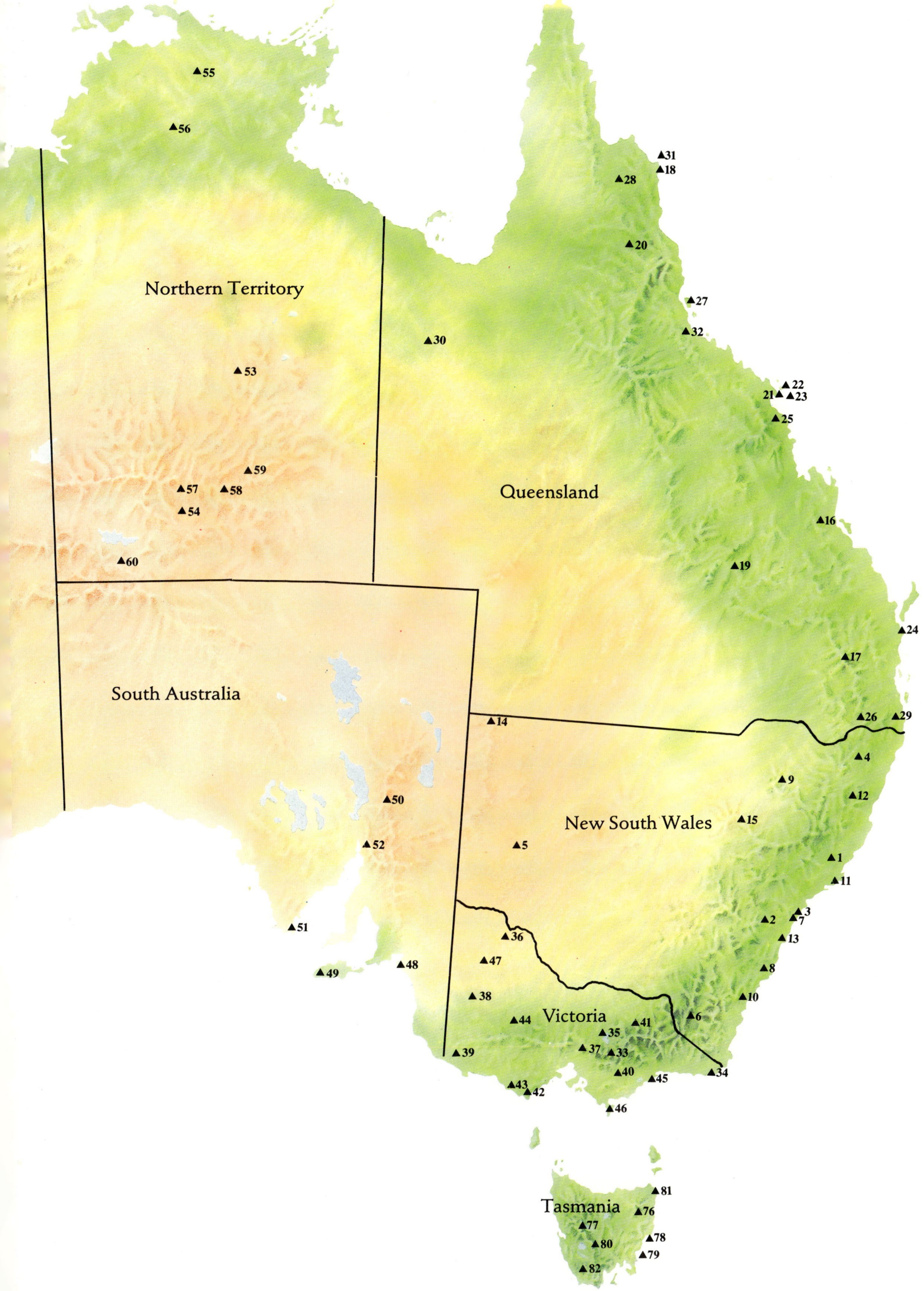

Northern Territory
Queensland
South Australia
New South Wales
Victoria
Tasmania
55
56
31
18
28
20
27
32
30
53
22
21
23
25
59
57
58
54
16
60
19
24
17
14
26
29
4
9
12
50
15
5
52
1
11
3
2
7
13
51
36
47
48
8
49
38
10
44
41
6
35
39
37
33
40
45
34
43
42
46
81
76
77
78
80
79
82

New South Wales

The many National Parks of New South Wales provide unspoiled scenic retreats, offering a wide range of activities for holidaymakers. The landscape here is surprisingly diverse; jagged volcanic spires tower over subtropical rainforests, wide flat plains and arid red sand dunes stretch as far as the eye can see, and coasts of cliffs stand sentinel over long sandy beaches.

Most of the state's parks are found along the coast, with their rugged headlands, quiet inlets and sweeping beaches pounded by the crashing surf. Sydney has many more hectares of parkland per head of population than any other metropolis in the world. The oldest parks in this area, the Royal and Ku-ring-gai Chase National Parks, provide agreeable settings for picnics and pleasant strolls, while further inland, the many splendid parks such as the mist-shrouded Blue Mountains National Park, give panoramic vistas of the surrounding countryside. Virtual wilderness areas offer limitless challenges to the adventurous at Kanangra Boyd, Wollemi, Marramarra and Dharug Parks.

North of Sydney, the forested plateau of Gibraltar Range National Park is home to the marsupial mouse, and various other marsupials find sanctuary in the deeply dissected ranges and valleys of Mount Kaputar National Park.

Kosciusko, embracing a large area of the continent's only extensive alpine region, is New South Wales' largest National Park. In the south-east of the state the steep ridges and ravines of Morton National Park throw down a gauntlet to the toughest bushwalkers, and oyster catchers and other wading birds can be seen in their thousands along the sheltered beaches and lake shores in the Murramarang National Park.

In the far west of New South Wales two outstanding National Parks are found. Kinchega houses beautiful lakes that provide an important breeding ground for a wide variety of waterbirds and Sturt National Park, the most remote park in the state, comprises open deserts and red sand dunes—a taste of the real Australian outback.

Left: *View of the magnificent waterways of Pittwater looking towards Barrenjoey headland from Ku-ring-gai Chase National Park.*

Barrington Tops
National Park

Straddling the Great Dividing Range, on the northern escarpment of the Hunter Valley where six major rivers spring into life, Barrington Tops National Park plunges from alpine plateaux to steaming rainforests. Such climatic variations support a remarkable and varied range of natural habitats rich in their diversity of plant and animal species.

Facilities

Car camping for conventional vehicles is limited to the Gloucester River camping area where there are toilets and fireplaces. The three campsites on Barrington Tops are accessible only by four-wheel-drive vehicles. There is a new camping ground near Barrington Guest House.

Bush camping is permitted anywhere in the park, but must be more than 300 m from roads, tracks, camping and picnic areas.

Picnic and barbecue areas are found in many locations in the park.

Accommodation is available at Barrington Guest House on the Dungog Road.

Location and Access

The park is 321 km north of Sydney and 38 km west of Gloucester. Access to the Gloucester River camping area is via the Gloucester Tops Road from the turnoff on the Gloucester–Stroud Road, 10 km south of Gloucester. This road is partly sealed and subject to flooding in very wet weather.

The northern area is reached via the unsealed road from Gloucester to Scone, and the southern arm via the partially sealed road from Dungog (43 km).

Trails within the park are suitable for four-wheel-drive vehicles only during dry weather. They are closed from 1 June to 30 September.

Landscape

The park covers 39 325 ha, and consists of two linked high altitude plateaux rising to 1590 m and falling steeply to 400 m. Numerous creeks and rivers flow either south to the Hunter River or north into the Manning River, forming raging torrents in the wet, and supporting majestic rainforests, swamps, bogs and fens. On the high plateaux are alpine woodlands, heavy with snow on cold winter days.

Flora and Fauna

Grey kangaroos, wallabies, pademelons, possums and gliders, bandicoots, wombats, native cats and marsupial mice inhabit the park.

Birds are everywhere in the forest, and include parrots and cockatoos, bowerbirds, brush turkeys, the bell miner, the whipbird and the superb lyrebird.

View across the densely timbered Barrington Tops National Park.

Moss-clad Antarctic beech trees thrive in the humidity of Barrington Tops.

Snow gums, grasslands and peat bogs characterise the vegetation of the high plateau. Lower down, rainforests are found side-by-side with wet and dry eucalypt forests, and Antarctic beech forests. Sassafras have taken over areas cleared by timber-getters who harvested the forests for red cedar trees. There are many epiphytic ferns, mosses and lichens, and stinging trees.

Walking Tracks
There are numerous walking tracks ranging from a tour of the Gloucester River camping area to invigorating hikes up to Mt Barrington. They follow creeks and rivers, pass swimming holes and picnic spots, through rainforests and woodlands to the high plateaux. The weather may change dramatically, so walkers should be prepared and advise friends of their planned route. Walking track brochures are available.

Other Activities
The park offers many opportunities for birdwatching and nature study. Horse riding, trout fishing, photography and bush camping are popular activities.

Climate
It can be very cold in the winter months although snowfalls on the plateaux are rare. Local conditions vary dramatically from alpine at the top with substantial rainfalls to temperate coastal weather on the lowlands. Visit all year round.

Bookings and fees
No bookings are accepted and no fees charged.

Information
The Superintendent
P.O. Box 270
Raymond Terrace
New South Wales 2324
Tel. (049) 87 3108

Blue Mountains
National Park

West of Sydney, the Blue Mountains rise up from the coastal plain to form a deeply etched, densely vegetated plateau. Precipitous cliffs provide breathtaking views across deep canyons, forests, woodlands and rainforest communities. This is the site of the Blue Mountains National Park and some of the most spectacular mountain scenery in Australia.

Facilities
In the Grose Valley area there is a small car camping site at Perry's Lookdown, with toilets, fireplaces, tank water and picnic tables. In the Glenbrook area there are car camping sites at Euroka Clearing, the Ironbarks, the Oaks, Murphy's Glen and Ingar, all with toilets, fireplaces and tank water. Caravans are allowed, but not encouraged. Creek water should be boiled.

*Grey butcherbirds (*Cracticus torquatus*) abound in Blue Mountains National Park.*

Bush camping is permitted at least 1 km from roads and facilities, but in the Upper Grose Valley and Grand Canyon you must use the designated campsites at Acacia Flat, Fortress Creek Flat and Burra Korain Flat. Camping is prohibited within 3 km of Lake Burragorang.

There are youth hostels at Katoomba and Hawkesbury Lookout, and visitor centres at Glenbrook, Blackheath and Waratah Native Gardens.

Picnic and barbecue areas are found in many locations in the park.

Location and Access

The park is approximately 100 km west of Sydney. Access is via the Great Western Highway, turn off at Glenbrook or take one of the side roads between Glenbrook and Blackheath; the Bells Line of Road between Windsor and Lithgow cuts across the park. A regular train service runs from Sydney to Mount Victoria.

Landscape

The park covers 215 955 ha of tall blue gum forests, heaths, mallee, swamp and grasslands, and is part of the central plateau of the Blue Mountains, forming the western wall of Sydney's Cumberland Basin. Rivers and creeks have cut through colourful sandstone producing great sheer-sided valleys and narrow canyons, particularly in the Grose River area. There are many spectacular waterfalls and the large Lake Burragorang which supplies Sydney's drinking water.

Flora and Fauna

Wallabies, possums, greater gliders, sugar gliders and marsupial mice are often seen.

More than 100 species of birds inhabit the park, they include lyrebirds, scrub wrens, whistlers, whipbirds, parrots, honeyeaters, finches, treecreepers and the occasional eagle.

Commonly encountered reptiles include goannas, jacky lizards, water dragons and skinks.

There are forests of great blue gums, rainforests with coachwood, lilly pilly, ferns and mosses, and springtime flower displays of waratahs, banksias, grevilleas and pea bushes.

Walking Tracks

Numerous well-maintained and signposted walking tracks cross the park, from easy nature trails along the cliff tops to week-long expeditions into remote areas. The Grose Valley and Glenbrook areas offer the most spectacular scenery and are most frequently visited. Here, there are forests, rivers,

Right: *The renowned wind-worn faces of the Three Sisters, remnants of an ancient river path.*

Left: *Breathtaking views of the Great Dividing Range from Pulpit Rock in Blue Mountains National Park.*

cliff edges and excellent bush camping areas. Red Hands Cave is an interesting Aboriginal site easily reached by foot.

Visitors are requested to complete an 'Intention Form' before climbing or overnight bushwalking in the park.

Other Activities

The cliffs offer many easy and challenging rock climbs. There are good fishing and swimming spots in the rivers.

Climate

Winters can be very cold. The weather can change dramatically at any time of year, and walkers should carry warm clothing. Visit all year round.

Bookings and Fees

Bookings are required for campsites in the Glenbrook area, where an entrance fee is payable.

Information and Bookings

National Parks and Wildlife Service
Blue Mountains District Office
Govetts Leap Road
Blackheath
P.O. Box 43 Blackheath
New South Wales 2785
Tel. (047) 87 8877, bookings (047) 39 2950

The beautiful Weeping Rock in Blue Mountains National Park.

The waratah, floral emblem of New South Wales, is one of the many glorious wildflowers seen in the Blue Mountains.

Left: *Cascades at Wentworth Falls in Blue Mountains National Park provide a delightful swimming spot for visitors.*

Bouddi
National Park

Wooded slopes tumble from the spine of Bouddi Peninsula to a choice of peaceful beaches and grass trees and burrawong cycads crowd the walking tracks that command panoramic views along the deeply indented coast of Bouddi National Park. Offering excellent opportunities for bushwalking all year round, Bouddi National Park has a magnificent coastline, home for a rich variety of plants and animals. The park also includes a marine extension of 283 ha of protected waters near Maitland Bay.

Facilities

The park offers facilities for day use, basic lightweight camping and some limited car-based camping.

Only the Putty Beach camp has vehicle access as well as toilets, water and barbecues. Tallow and Little Beach campsites, ranked by swimmers and anglers as the most pleasant in the state, have toilets, barbecues and freshwater creeks (it is advisable to boil the creek water before drinking). Day-use sites include Mt Bouddi, Lobster Beach and Maitland Bay.

Location and Access

Bouddi National Park is situated about 20 km south-east of Gosford on the Scenic Road through Kincumber, or via Empire Bay Road and Wards Hill Road from Ettalong. From the Scenic Road there are five points of entry into the park.

Landscape

Situated along a narrow coastal strip on the northern entrance to Broken Bay, Bouddi National Park covers 1147 ha of coastal heath and dry sclerophyll forest. The horizontally bedded shales, sandstones and conglomerates found here are about 180–200 million years old. A small coastal range capped by Hawkesbury sandstone runs parallel to the coastline with the highest point being 152 m above sea level at Mt Bouddi.

Flora and Fauna

There is rich diversity of vegetation in the park. Behind Maitland Bay, deep valleys support a coastal rainforest of bangalow, cabbage tree palms and coachwood. Blackbutt and mahogany are found on the shales in many gullies and the sandstone slopes are clad in fine stands of eucalypts. Angorphora species also occur with an understorey of banksia, acacia, burrawong palms and grass trees. The coastal heaths come alive in spring with a profusion of wildflowers.

Despite the varied vegetation, few large animals remain in the park, due mainly to the limited area and adjacent settlements. However, spiny anteaters, bandicoots, antechinus, possums, gliders, and swamp wallabies make their home here and reptiles are common. Bird life and insects are the most obvious forms of wildlife, and extensive rock platforms support a large variety of marine life.

Walking Tracks

An extensive network of pleasant walking tracks may be followed thoughout the park to beaches and lookouts with magnificent views of the coastline, Brisbane Waters and the hinterland.

Waves crashing on Little Beach in Bouddi National Park.

Other Activities

Swimming, fishing and snorkelling are popular among holidaymakers, while the beautiful coastline and forest scenery offer opportunities for bushwalking and photography. The remains of the paddlewheeler, SS *Maitland Bay,* wrecked here with the loss of 26 lives in 1898, can still be seen on the rocky headland of Maitland Bay.

Climate

The park has a temperate coastal climate which can be enjoyed all year round. Visit in spring for the wildflowers.

Bookings and Fees

Fees are charged for camping and bookings are required. A ballot may be required for periods of highest demand.

Information and Bookings

National Parks and Wildlife Service
Hawkesbury District
P.O. Box 1393 Gosford South
New South Wales 2250
Tel. (043) 24 4911

Gibraltar Range

National Park

On a high escarpment at the eastern edge of the New England plateau, the Gibraltar Range National Park provides the first breath of cool mountain air after the summer heat of the coastal plains of north-eastern New South Wales. A place steeped in legends of the Aboriginal Dreamtime, the home of the scarlet waratah and the rufous rat kangaroo, this is an area of deep gorges, granite tors and tumbling waterfalls.

Facilities

Caravan and car camping sites are located at Mulligan's Camp, where there are fireplaces, pit toilets and showers. Bush camping is allowed anywhere in the park, but must be more than 1 km from roads and facilities.

There are a number of picnic areas in the park and a visitor centre at the Dandahra Creek entrance to Mulligan's Road.

A spectacular waterfall plunges through the deeply dissected plateau of Gibraltar Range.

Location and Access

The park is 672 km north of Sydney, 79 km from Grafton and 69 km from Glen Innes. The Gwydir Highway runs through the park, turn off on to the narrow, unsealed Mulligan's Road to the camping area (10 km).

Landscape

The park covers an area of 17 273 ha on an elevated plateau some 1200 m above sea level. Deep gorges dissect the plateau, many with spectacular waterfalls plunging more than 200 m into the steep-sided valleys. Huge granite tors with names like the Old Man's Hat, the Anvil Rock, the Haystack and the Waratah stud the skyline.

Flora and Fauna

A number of rare animal species live in the park, they include the rufous rat kangaroo, parma wallaby and the spotted quoll. Parma wallabies, pademelons, possums and gliders, and marsupial mice are occasionally seen, and a platypus may be spotted around one of the creeks.

Some 223 species of birds have been recorded in the park. The rufous scrub bird is abundant here, although rare elsewhere; there are lyrebirds, bell birds, whipbirds, bowerbirds, rosellas, parrots, honeyeaters and a number of wrens.

In the west are great forest eucalypts, mountain ash, manna gum and blackwood. In the gorges are rainforest areas with cedar, rosewood, native ash, stinging trees, strangler figs and coachwoods. The swamps on the plateau are perfect for Christmas bells. There are 4000 ha of waratahs around Guangan's Rock, and many orchids bloom on the forest floor and in the shade of the boulder outcrops. Waratahs are found in all parts of the park, except in the rainforest. They flower during late October and early November.

Early morning mists are frequent in the Gibraltar Range and visitors should be prepared for sudden cold snaps at any time of the year.

Left: *The Needles, a striking granite rock formation in Gibraltar Range National Park.*

Walking Tracks

Several walking trails begin at Mulligan's Hut and the visitor centre, varying from 1–10 km, through rainforests, past creeks, pools, waterfalls and lookouts. All are well marked and maintained.

Other Activities

There are some excellent rock pools for swimming, some of the creeks are suitable for canoeing, but beware of waterfalls. The cliff faces and tors offer challenges for rock climbers, and the wide and usual range of flora and fauna provide many opportunities for nature study and photography.

Climate

Summers are warm and wet and can turn suddenly cold, winters are much drier but cold at night. Visit all year round.

Bookings and Fees

Bookings are not required and no fees are payable.

Information

Officer-in-charge
National Parks and Wildlife Service
Glen Innes District
P.O. Box 281
Glen Innes
New South Wales 2370
Tel. (067) 32 1177

Kinchega

National Park

In the far west of New South Wales, along the banks of the Darling River, more than 1000 km from its source in the Great Dividing Range, lies Kinchega National Park. This is an area of red sand dunes and black soil flats alongside the winding river, its flood channels and overflow lakes. The park is a birdwatcher's paradise, and offers a fascinating glimpse of a huge early pastoral settlement.

Facilities

There are car camping sites along the western bank of the Darling River with fireplaces and access to water, toilets and hot showers. A number of picnic areas have been established along the lake shores, and the ranger station is at Emu Lake. Bush camping is permitted, but the ranger should be informed.

Accommodation is available at Menindee.

A cormorant, one of the many water birds found around the Darling River in Kinchega National Park.

Flora and Fauna

The open plains are home to red kangaroos and emus. In the woodlands grey kangaroos, euros and possums are common.

The bird life is prolific, and of the 374 species recorded, 65 are found in lake or river habitats. Here are pelicans, black swans, ducks, egrets, cormorants, seagulls, stilts and avocets, while parrots are common in the woodlands and hawks and wedge-tailed eagles can be seen circling in the sky.

The floodplains support mainly black box trees, while river red gums and coolabahs grow along the river banks. On the plains are mainly bluebush and prickly wattle, yet after rain the landscape is transformed into a carpet of grasses and paper daisies.

Walking Tracks

There are well-marked lake edge waterbird walks and tracks along the river banks, offering many opportunities for closer observation of the waterbirds and kangaroos, and the remains of Aboriginal campsites.

Other Activities

There are a number of interesting drives through the park revealing the history and nature of the area. A large and impressive shearing shed, built in the 1880s and part of the sprawling Kinchega Station, is maintained as a museum in the park. Swimming, boating and photography are popular activities.

Climate

The area has an average annual rainfall of only 216 mm, with very hot summers. Visit in winter and early spring.

Bookings and Fees

No bookings are accepted or fees payable.

Information

The Senior Ranger
Kinchega National Park
P.O. Box 40
Menindee
New South Wales 2879
Tel. Menindee 14

Location and Access

The park is 1018 km west of Sydney and 111 km south-east of Broken Hill. Access is via the sealed road from Broken Hill, via the partly sealed road from Wentworth in the south, or along the unsealed road from Ivanhoe to the east. Menindee, 2 km west of the park, is on the Indian Pacific railway line. Roads within the park are unsealed.

Landscape

The park covers 44 182 ha and is dominated by the meandering Darling River and two of its saucer-shaped overflow lakes where excess floodwater is stored. Black soil flood plains covered with trees border the river, while most of the park consists of red sandy soil supporting dry grasses. Crescent-shaped sand dunes known as lunettes occur on the north-eastern shores of the lakes.

Top right: *Flowering dock blooms across the arid Kinchega plains after rain.*

Bottom right: *Emus are often seen on the grassy woodlands of the Darling River plain.*

Far right: *The Darling River at Kinchega National Park, fringed by river red gums and coolibahs.*

Kosciusko
National Park

Right in the heart of the Snowy Mountains, Kosciusko National Park is a place of vast alpine snowfields, glacial lakes, forested gorges and grassy plains carpeted with beautiful wildflowers in summer. Australia's highest mountains are found here, and the winter snow and summer rains give rise to some of the greatest rivers in the country.

Facilities
Visitor centres are located at Sawpit Creek and Khancoban. Camping along the roadsides is permitted at rest areas with basic amenities: pit toilets, fireplaces and a water supply.

Bush camping is allowed almost anywhere in the park, but must be at least 1 km from roads and facilities, and more than 100 m from the glacial lakes. Above the tree-line only portable gas or spirit stoves may be used.

There are picnic areas with fireplaces and toilets in many locations throughout the park.

Accommodation is available in lodges and hotels in the park and nearby towns.

Right: *An approaching storm looms over the snow-covered landscape at Kosciusko National Park.*

Frozen branches and countryside give a fairytale atmosphere to winter time in Kosciusko.

A Willie wagtail (Rhipidura leucophrys), *feeds its young in Kosciusko National Park.*

Thousands of winter visitors flock to the skifields at Kosciusko. Here a chairlift at Smiggin Holes carries skiers to the mountain heights.

Location and Access

The park is about 490 km south-west of Sydney and 200 km south-west of Canberra. Access is via the Snowy Mountains Highway from Tumut to the north or Cooma in the south-east; from Jindabyne via the Alpine Way or Kosciusko Road; and from the west via Khancoban. The major access roads within the park are sealed, although there may be freak snowfalls in summer on the high passes. Snow may close roads from May to October, and snow chains should be carried. The Khancoban road to Cabramurra is closed in winter. Check local road conditions before travelling in winter.

Landscape

The park, which covers 690 000 ha, is the largest in New South Wales and contains Australia's greatest alpine area. With a range of altitude varying from 200 m to more than 2100 m, the park has an unusual variety of flora and fauna, and boasts the highest mountains in Australia, with Mt Kosciusko reaching 2229 m.

The landform is one of undulating hills and rounded mountain peaks falling off on the western side to form deep gorges and precipitous slopes. Snowbound granite outcrops contrast with blue lakes and mountain streams, fern gullies and giant trees. The park is the source of such great rivers as the Murray, Snowy and Murrumbidgee.

Flora and Fauna

Large mammals include the grey kangaroo, red-necked wallaby and the wombat. Possums, echidna and bush rats are common, and the alpine areas are home to the rare mountain pygmy possum which lives under the snow in winter.

Falcons, kestrels and ravens are common in the alpine regions. There are magpies, cockatoos, rosellas, honeyeaters, treecreepers, the bright-breasted flame robin and occasional thornbills to be seen.

The vegetation varies from alpine herbfields with silver snow daisies, billy buttons and eyebrights, through forests of snow gums to fern gullies. The trigger plant, heath myrtle, swamp heath, everlasting daisies and alpine sundew are often seen below the tree line.

There are a number of snakes in the lower regions including the copperhead, tiger and brown snake. Native trout occupy some of the streams and pools, although the introduced rainbow and brown trout are more common.

Walking Tracks

The park offers many interesting walks varying from easy nature trails to long walks across country, around the lakes and through forests to the mountain peaks. Weather conditions are very variable, and bushwalkers should be equipped with strong footwear, wet weather gear, warm clothing, emergency rations and medical supplies. They are requested to register at one of the ranger stations before leaving.

Left: *Charlotte Pass, a popular ski resort in Kosciusko National Park.*

Other Activities

The park is Australia's main skiing centre in winter. Both downhill and cross-country skiing are very popular. There are extensive, spectacular limestone caves in the Yarrangobilly Caves area open for inspection, alone or with a guide. Nearby Kiandra is an old goldmining town with an interesting historical collection.

Swimming is possible in many of the creeks and in the thermal pool at Yarrangobilly. There are several good rivers for canoeing and good trout fishing in many streams and lakes (licence required). Other popular activities are climbing, cycling and horse riding, although the latter is restricted to certain areas and the route must be checked with the ranger before leaving.

Climate

The park has alpine weather conditions. These are very variable, and can change dramatically in the space of a few hours, ranging from hot summer days to freezing nights and heavy snowfalls. Visit in spring and summer for beautiful wildflower displays, and in the winter for skiing.

Bookings and Fees

Fees are payable on entry to the park.

Information

The Superintendent
Kosciusko National Park
Private Mail Bag
Via Cooma
New South Wales 2630
Tel. (064) 56 2102, camping and caravans (064) 56 2224

Sunlight glistening on the frozen landscape is a photographer's paradise in Kosciusko National Park.

Ku-ring-gai Chase
National Park

Right on the northern boundary of Sydney, in a coastal setting unequalled in Australia, lies one of the country's oldest and most magnificent National Parks. Ku-ring-gai Chase National Park is a place of sublime views over cracked and rugged sandstone plateaux to the still, blue waters and secluded beaches of Pittwater, Broken Bay, the Hawkesbury River and its tributaries.

Facilities
Camping is only allowed at The Basin camping area on Pittwater, which can be reached by boat, ferry or by walking from West Head Road. There are toilets, cold showers and fireplaces, although visitors should bring their own cooking fuel. Booking is essential as the area is very congested during peak holiday periods.

There are kiosks at Bobbin Head, Appletree Bay, Akuna Bay and Cottage Point. A youth hostel is located at Towlers Bay, and a visitor centre at Kalkari.

Picnic areas, toilets and fireplaces are found in many locations. There are boat ramps at Appletree Bay and Akuna Bay, boats can be hired at Bobbin Head, Akuna Bay and Cottage Point; fuel and bait are also available.

Location and Access
The park is 30 km north of Sydney. Access is via the Pacific Highway, turn off at Pymble or Mt Colah for Bobbin Head; or via the Mona Vale Road, turn off at Terrey Hills for West Head.

A regular ferry boat runs from Palm Beach to The Basin, there are railway stations at Turramurra, Mt Ku-ring-gai and Berowra, and a bus service from Turramurra Station to Bobbin Head.

The major access roads within the park are sealed.

Landscape
The park covers 14 717 ha of rugged sandstone country set around the flooded river valley of Broken Bay. Deep, narrow valleys cut through the sandstone by Cowan Creek and other watercourses have formed sheltered waterways long used by sailors and anglers. There are many beautiful secluded bays, woodlands, heaths and an amazing variety of plants growing on the poor soils of the area.

Flora and Fauna
Swamp wallabies, wallaroos, possums and gliders, koalas, bandicoots, marsupial mice, echidna, tortoises, goannas and snakes are occasionally seen.

More than 180 species of birds have been recorded in the park, they include the lyrebird (found at West Head and Garigal in winter), the kookaburra, parrots, honeyeaters eagles and many waterbirds.

Forty fish species have been recorded in the extensive tidal waterways. They include sharks, bream, blackfish, mullet, hairtail jewfish and the leatherjacket.

Wildflower displays colour the landscape in winter and spring with the heath banksia attracting hundreds of honeyeaters. Among the tree species are found many eucalypts, angophoras, she-oaks and wattles. Shrubs include banksias, grevilleas, waratah, hakeas, epacris, boronias, pea flowers, guinea flowers, flannel flowers and grass trees.

Walking Tracks
There are 17 signposted walking tracks on West Head and seven on Bobbin Head. They vary from easy strolls with excellent views to day-long hikes past Aboriginal engravings, through forests and heathlands to spectacular lookouts and bays. A nature track at Kalkari introduces visitors to the bushland.

Other Activities
Boating is popular and an excellent way to visit the park. There are good fishing spots and many places to swim, with shark nets at Appletree Bay and The Basin. There are hundreds of Aboriginal engravings in the park and many middens, providing a reservoir for research.

Left: *Aboriginal carvings on rock faces in Ku-ring-gai Chase National Park.*

Right: *Charming Cottage Point is a popular anchorage in Ku-ring-gai Chase National Park.*

View from West Head in Ku-ring-gai Chase National Park towards Barrenjoey Point at Palm Beach, north of Sydney.

Right: *The flannel flower, part of the colourful winter-spring wildflower display at Ku-ring-gai Chase.*

Horse riding is a popular activity, and there are a number of authorised trails in the park.

The visitor centre presents an audio-visual display and many animal exhibits. The park offers good opportunities for birdwatching, nature study and photography.

Climate

Like Sydney, the park has warm summers and cool winters, with the highest rainfall in the summer months. Visit all year round.

Bookings and Fees

Bookings must be made for the campsite. Park entrance and camping fees are payable.

Information and Bookings

The Superintendent
Ku-ring-gai Chase National Park
Bobbin Head, Turramurra
New South Wales 2074
Tel. (02) 457 9322 or 457 9853, the Basin camping area (02) 919 4036

Morton
National Park

South-west of Sydney, just 30 km from the coast, the Bundawang Range rises from the plain in a series of great gorges and steep-sided valleys. This richly vegetated, rugged wilderness area of majestic waterfalls and massive cliffs is the site of Morton National Park, one of the large National Parks in the state.

Facilities

There are caravan and car camping areas at Fitzroy Falls, Gambells Rest and Bundanoon with fireplaces, toilets and showers. Bush camping is allowed anywhere in the park, but must be at least 1 km from roads and picnic areas.

The visitor centre is at Fitzroy Falls, and there are picnic facilities with fireplaces in many locations in the park.

There is a youth hostel at Bundanoon and accommodation at Moss Vale.

The satin bowerbird shows a preference for blue objects when building its nest in Morton National Park.

A striated pardalote (Pardalotus striatus) *or 'chip chip'*

Location and Access

The park is 155 km south-west of Sydney and 17 km south of Moss Vale. Access is via the Moss Vale Nowra road to Fitzroy Falls and Bundanoon, to Tianjara Falls via the unsealed road from Nowra to Braidwood, and to the southern section of the park via the Princes Highway, 16 km south of Ulladulla.

Landscape

The park covers 152 948 ha of rugged sandstone eroded into deep gorges by the waters of the Shoalhaven, Kangaroo, Endrick and Clyde River systems. The terrain is one of colourful cliff faces from which waterfalls plunge into the densely forested valleys below. The heath-covered plateau provides spectacular views, while the southern areas offer vast tracts of undisturbed countryside.

Flora and Fauna

Grey kangaroos, swamp wallabies, red-necked wallabies, possums, wombats, bandicoots, goannas, lizards and snakes are occasionally seen.

The varied bird life includes large birds of prey such as wedge-tailed eagles and whistling kites. Woodland species include lyrebirds, parrots, cockatoos, rosellas, kookaburras, treecreepers, scrub wrens and bowerbirds. In the heaths are two threatened species: the eastern bristle bird and the swamp parrot.

On the plateau a variety of trees are found, they include black ash, scribbly gum, Sydney peppermint, old man banksia and thin-leaved stringybark. In the rainforests are giant red cedar and coachwood trees and many tree ferns. The great variety of wildflowers include crimson bottlebrushes, trigger plants, red honey flowers and the broad-leaved hakea.

Walking Tracks
There are short, easy walks to lookouts, nature trails and a number of scenic walking tracks around the escarpment with excellent views over the valleys and falls. Steep tracks descend to the creeks, and there are opportunities for long bushwalks in isolated countryside.

Climate
The park has a temperate coastal climate. Visit all year round.

Bookings and Fees
Bookings are required for Bundanoon camping areas. Sites at Fitzroy Falls cannot be reserved. Camping fees are charged.

Information and Bookings
National Parks and Wildlife Service
Fitzroy Falls Information Centre
Nowra Road
Morton National Park
New South Wales 2577
Tel. (048) 87 7270

The jutting peaks of the Nandewar Range in Mt Kaputar National Park.

Mount Kaputar National Park

Some 300 km due west of Coffs Harbour, Mt Kaputar rises majestically above the great inland plains. Part of the Nandewar Range, this ancient volcanic area is a place of forested mountains, waterfalls and fern gullies; a place where the eagles glide in the cool mountain air; a place of endless sunsets over the dry heartlands of Australia.

Facilities
There are two campsites with vehicular access. Dawsons Spring at 1373 m has two cabins, campsites, fireplaces, toilets, hot showers, a laundry, visitor centre and picnic area. The cabins accommodate up to two adults and four children or four adults only, and are equipped with all facilities. Bark Hut camping area has fireplaces, toilets and hot showers. Caravans are not allowed in the park.

Bush camping is permitted in the park, but must be more than 1 km from roads and facilities, and the ranger should be notified.

Location and Access
The park is 570 km north-west of Sydney and 53 km east of Narrabri. Access is from Narrabri to Dawsons Spring via a steep and narrow gravel road, or via the gravelled Narrabri–Bingara road to Killarney Gap.

Landscape
The park covers 36 867 ha, and follows the main ridge of the Nandewar Range, the remnants of a large volcanic centre active some 18 million years ago. Mt Kaputar rises to 1508 m and lies on the rim of an ancient volcanic cone. This and several smaller peaks nearby form a skyline of jagged mountains falling away in forested slopes and deeply cut valleys to the plains below. Here are steep cliff faces and the barren rocks of old lava flows, cascading rivers, clear streams and fern-filled gullies.

Flora and Fauna
The park is an island sanctuary for many mammals isolated by the surrounding cleared agricultural lands. Here you will find grey kangaroos, wallaroos, koalas, possums, greater gliders, feathertail gliders and marsupial mice.

Whistling and wedge-tailed eagles are often seen riding the thermals, whereas in the forests are brush turkeys, parrots, crimson rosellas, treecreepers, honeyeaters and many more bird species.

Snow gums are found in majestic stands on the peaks, lower down are forests with white box, black cypress, red stringybark, grass trees, native cherry trees and patches of wet sclerophyll forest with a rich understorey of ferns. Acacias and the many wildflowers bloom at different times throughout the year.

Walking Tracks
A range of walking tracks varying from an easy and rewarding nature trail to strenuous full-day hikes have been constructed. They encompass forests and cool snow gum wood-

Swan Rocks, a peculiar formation in Mt Kaputar National Park resulting from ancient volcanic activity.

lands, views across the jagged peaks and deep gorges to clear streams and beautiful waterfalls. Water should be carried on the longer walks; sturdy footwear is essential, wet-weather gear and warm clothing may be required.

Other Activities
There are good rock climbing sites, although the ranger should be consulted, and many opportunities for geological and nature study and photography.

Climate
The park has hot summers and cool winters. Visit all year round.

Bookings and Fees
Bookings are required for the cabins. Fees are payable for the cabins and campsites at Dawsons Spring.

Information and Bookings
National Parks and Wildlife Service
Narrabri District Office
P.O. Box 72
Narrabri
New South Wales 2390
Tel. (067) 92 7222

Murramarang
National Park

Below the steep slopes of the Murramarang Range on the south coast of New South Wales, a narrow belt of shoreline about 27 km long is the site for the Murramarang National Park. Headlands, cliffs, beaches, wave-cut platforms and small offshore stacks and islands provide a diversity of coastal landforms. The vegetation supporting a variety of bird and animal life ranges from tall eucalypt forests and woodland through to the flowering heaths, banksia and casuarina scrubs of the headlands and beaches. Murramarang includes four nearshore islands, Wasp Island, Dawsons Island, O'Hara Island and Grasshopper Island. The mainland section of the park is divided into two parts separated by the towns of Durras North and South Durras.

A rugged section of coastline north of Batemans Bay at Murramarang National Park.

Facilities

The southern section of Murramarang is serviced with camping grounds as well as toilets, showers and barbecue facilities near South Durras village on the southern shores of Beagle Bay. Durras North has three camping areas. Together with the cabins, camping ground and barbecue facilities at Pebbly and Depot Beaches these cater for the central portion of the park. Camping amenities are also provided on the northern tip at Pretty Beach.

Bush camping is permitted but must be at least 100 m from picnic areas, roads and walking tracks while camping on the beaches and headlands is prohibited due to the fragile nature of the coastal ecology.

Nearby Batemans Bay and Ulladulla provide hotels, motels and other camping areas.

Location and Access

Murramarang is located 280 km south of Sydney and 10 km north of Batemans Bay. Access is via the Princes Highway, north of Batemans Bay, turning off at East Lynne for Durras North, Pebbly Beach and Depot Beach. The northern tip of the park is reached via Kiopa. In order to reach South Durras and the southern section of the park, motorists must leave the highway at Benandera.

Landscape

The park, covering 1597 ha, is barely 2 km across at its widest point. It includes the coast and coastal ranges from near Kiopa in the north to North Head at Batemans Bay in the south. Eastern spurs of the partly drowned edge of the Murramarang Range remain as rocky headlands and shore platforms jutting into the sea.

Around Snapper Point in the north, fossil-bearing rocks are featured while there is the distinctive Sydney sedimentary basin on the southern boundary outcrops near Dark Beach.

Over 20 sites in the park have been recorded as showing evidence of a long period of Aboriginal culture in the region.

Interesting rock formations weathered by wind erosion and ocean spray in Murramarang National Park.

Flora and Fauna

The vegetation in the park is dominated by tall spotted gums with a burrawang palm understorey. Around Durras Mountain some isolated gullies have small rainforest communities. Closer to the coast, lower growing trees and shrubs such as geebungs and casuarina are abundant.

Eastern grey kangaroos and wallabies are particularly common sights at dawn and dusk when they emerge from cover to feed. Fascinating shell animals are to be seen in the rock pools of the seashore while oyster catchers and other wading birds roam along the sheltered beaches and lake shores. Parrots, honeyeaters, finches and thrushes are attracted to the forest and flowering shrubs.

Quiet sheltered beaches at Murramarang National Park are popular for swimming, fishing, skindiving and camping.

Walking Tracks
In the southern section of Murramarang there are a number of walking tracks leading down to the beach and headlands. Richmond Headland offers panoramic views of the coastline and glorious wildflowers from July through to October.

In the north, tracks may be followed through forest and woodland to the coast. The steep climb to the peak of Durras Mountain is rewarded by panoramic views of the tablelands and coastline.

Other Activities
Beach and rock fishing, swimming and surfing are popular among holidaymakers while the beautiful coastline and forest scenery offers opportunities for bushwalking and photography. The unusual coastal landforms and fossil-bearing rocks also provide an interesting focus for nature study.

Climate
The park has a temperate coastal climate which may be enjoyed year round.

Bookings and Fees
Camping or van sites are often difficult to obtain in holiday seasons so bookings are advisable. Park entrance and camping fees are payable. Bookings may be made directly with the various caravan parks and camping areas in Murramarang National Park.

Information and Bookings
National Parks and Wildlife Service
Nowra District
24 Berry Street
Nowra
New South Wales 2540
Tel. (044) 21 9969

Myall Lakes
National Park

A place of peace and solitude in the centre of the busy east coast of New South Wales, Myall Lakes National Park is one of the least spoiled and tranquil coastal areas in the state. A system of waterways covering more than 10 000 ha and separated by high sand dunes from the ocean beaches offers relaxation and the enjoyment of a rich and varied flora and fauna.

Right: *Tranquil sunset over the Myall Lakes.*

Houseboats are a pleasant way to enjoy the seclusion and charm of Myall Lakes National Park.

Facilities
Caravan and car camping sites are located at Legge's Camp at Bombah Point, where there are fireplaces, toilets, hot showers, a laundry, kiosk and boat fuel. Booking is essential during the busy holiday periods. Car camping and caravans are also permitted at Mungo Brush and Violet Hill where there are fireplaces and pit toilets only. Other boat access campsites occur around the lake.

Bush camping is permitted, but must be more than one kilometre from roads and facilities. There is a picnic area at Mungo. Drinking water is scarce and should be carried in.

Accommodation is available at Bulahdelah, Forster and Hawks Nest.

Location and Access
The park is 236 km north of Sydney and 16 km east of Bulahdelah. Access is via the Pacific Highway, turn off to Tea Gardens and from there follow the unsealed road north along the coast, or turn off at Bulahdelah on the unsealed Lakes Road, or turn off on the Lakes Way north of Bulahdelah to Bungwahl and follow the unsealed Seal Rocks Road. Roads within the park are unsealed. A public ferry links Bombah Point to the Mungo Brush Road.

Landscape
The park covers 31 005 ha, and is dominated by a series of large, natural, fresh-to-brackish lakes linked to the ocean by the Myall River. The lakes are fringed with forests and separated from the ocean by windswept sand dunes covered in parts by forests of tall eucalypts. The long, deserted beaches are broken by spectacular rocky headlands; offshore lies Broughton Island.

Flora and Fauna
Mammals found in the park include grey kangaroos, swamp and red-necked wallabies, possums and gliders, bandicoots, echidnas and marsupial mice. Many waterbirds are attracted to the lakes, ducks, cranes and black swans are common, while white-breasted sea eagles glide overhead looking for fish. The lakes are breeding grounds for prawns and many species of fish.

Pine tree plantations along the Myall River give way to stands of paperbarks and *Angophora costata* bordering the lakes, interspersed with rainforest species such as cabbage tree palms. The dunes are thickly forested with eucalypts and wattles in places, and wildflower displays brighten the coastal heathlands in spring.

Walking Tracks
There are a number of bushwalking tracks leading through the forests down to the lakeside or to the ocean beaches.

Other Activities
The best way to see the park is by boat. Small motor boats and houseboats can be hired at Tea Gardens, and at Bulahdelah there are boat launching ramps in the park. The lakes are perfect for canoeing, sailing, swimming and fishing. There are good surfing and fishing spots off the beaches.

Climate
The park has a temperate coastal climate. Visit all year round.

Bookings and Fees
Fees are payable at the camping areas throughout the parks and for boats moored off these areas. Bookings are only required at Legge's Camp.

Information and Bookings
The Superintendent
National Parks and Wildlife Services
P.O. Box 270 Raymond Terrace
New South Wales 2324
Tel. (049) 87 3108

Legge's Camp Bookings
Legge's Camp
Via Bulahdelah
New South Wales 2423
Tel. (049) 97 4495

New England National Park

High on the precipitous eastern edge of the Great Dividing Range, the New England National Park looks down over rugged ridges and densely forested, untouched valleys to the deep blue of the Pacific Ocean. A place of swirling mists and breathtaking views, the park offers a multitude of habitats from subalpine to subtropical, with a vast and varied range of wildlife.

Facilities
Caravans are allowed at the Styx River rest area, 2 km before the park entrance, where there are toilets, barbecues and running water. There is a car camping area at Thungutti, within the park, with toilets, showers and barbecues.

Bush camping is allowed anywhere in the park, but must be a least 1 kilometre from roads and picnic areas, and the ranger must be notified.

At Banksia Point is the Residence with accommodation for ten people and the Chalet with accommodation for six people with all facilities. Tom's Cabin has accommodation for eight with basic facilities.

There is a picnic shelter and fireplace at Point Lookout, and a number of picnic areas with barbecues and toilets in the park.

There is accommodation at Ebor and Dorrigo.

Location and Access
The park is 576 km north-west of Sydney, 85 km east of Armidale and 135 km west of Coffs Harbour. Access is via the Armidale–Ebor–Grafton road, turn off 9 km south of Ebor. Roads within the park are mainly unsealed.

Sunrise over cloud enshrouded New England National Park.

Landscape

The park covers 29 881 ha and extends along the eastern edge of a high plateau from which blunt basalt peninsulas project. Precipitous cliffs descend from 1566 m at Point Lookout to a deep, folded valley region, densely forested, with many virtually inaccessible peaks and ridges. Here the Bellingen, Nambucca and other rivers have carved their courses, draining the water from the often mist-enshrouded plateau. Beautiful temperate and subtropical rainforests fill the valley floors, while heath and snow gums grow on the cliff tops.

Flora and Fauna

Grey kangaroos, wallaroos, red-necked wallabies, swamp wallabies, possums and gliders, and the dusky marsupial mouse inhabit the park.

Of the 110 bird species recorded, 20 belong to the subtropical rainforest. Superb lyrebirds, scrub turkeys and cockatoos range widely. There are many pigeon species, currawongs, honeyeaters, robins, and the monotonous calls of cuckoos are often heard.

Common reptiles include goannas, the blue-tongued lizard, bearded lizard, water dragons, leaf-tailed gecko and the red-bellied black snake. There are many frogs, including the sphagnum frog.

The high plateau supports snow gums, stands of Antarctic beech and heathlands. Eucalypt forests clothe the valley sides while dense rainforests with red cedar, booyong, carabeen, staghorns, treeferns and orchids follow the creeks and rivers. There are excellent wildflower displays in spring.

Walking Tracks

There are 20 km of constructed walking tracks, including the spectacular Lyrebird Trail which follows the edge of the escarpment. There are many challenging walks for experienced bushwalkers in the trackless wilderness of the Bellingen River area. The park rangers will offer help and advice.

Climate

The plateau has mild, wet summers and cold winters with occasional snow. Warm clothing and rain-wear should always be carried. Visit all year round.

Bookings and Fees

Camping is free, and no bookings are accepted. The Residence, Chalet and Tom's Cabin must be booked and fees paid.

Information and Bookings

National Parks and Wildlife Service
Dorrigo District Office
P.O. Box 170
Dorrigo
New South Wales 2453
Tel. (066) 57 2309

View from Point Lookout (1566 m) over the densely forested peaks and valleys of New England National Park.

Ferns and mosses flourish in the cool damp gorges of New England National Park.

Royal
National Park

On the southern border of Sydney, where rugged cliffs are pounded by the restless Pacific, where quiet freshwater pools support pockets of rainforest, and heaths and woodlands bloom with countless wildflowers, lies Australia's first National Park. A wonderland of nature, right on the doorstep of the country's largest city.

Facilities
Caravan and car camping are allowed only at Bonnie Vale, where there are showers and toilets. Bookings are essential, but during peak holiday periods a ballot system is used. There are shops at Mainabar and Bundeena, 2 km from the camping area, and kiosks at Audley, Wattamolla and Garie.

In summer time hundreds of visitors enjoy the clear blue waters at the ocean entrance to Wattamolla Creek.

Wattamolla Falls cascade into a deep pool, popular with swimmers at the Royal National Park.

Bush camping is permitted in the park, but permits must be obtained. Check at the visitor centre for camping locations. No wood fires are permitted.

There is a youth hostel at Garie Beach. For information contact the Youth Hostel Association in Sydney.

A visitor centre is located at Audley, approximately 2 km from the Royal National Park railway station.

Location and Access

The park is 36 km south of Sydney and 47 km north of Wollongong. Access is via the Princes Highway; turn off past Sutherland for Audley, or at Waterfall; from Wollongong via Stanwell Park. A regular ferry service runs from Bundeena to Cronulla, and there is a rail link to Sydney and Wollongong, with a station in the park and others alongside the park's western boundary where walking tracks begin.

The major access roads within the park are sealed.

Landscape

The park covers 15 014 ha of mainly heath-covered sandstone, shaped like a saucer rim descending from 300 m in the south to the tidal estuary of Port Hacking in the north. Deep gorges have been cut through the sandstone plateau by the Hacking River and other watercourses. Creeks, swamps, waterfalls, clear pools, eroded sandstone cliffs, beautiful isolated beaches, rainforest pockets and woodlands characterise the landscape.

Flora and Fauna

Swamp wallabies, ringtail and brushtail possums, bush and swamp rats, the New Holland mouse, the antechinus, echidna and deer inhabit the park, but most are nocturnal.

More than 200 species of birds have been identified, the more commonly encountered ones include kookaburras, rosellas, cockatoos, ravens, currawongs, sea eagles, honeyeaters, wattlebirds, lyrebirds, green catbirds and satin bowerbirds.

Early morning, overlooking peaceful Wattamolla Creek.

Yellow-tufted honeyeater in Royal National Park.

Snakes are occasionally encountered where the track follows the creek, and funnel-web spiders are found in the park. Whales are seen offshore at some times of the year.

There are spectacular wildflower displays in the spring, particularly on the heaths, with mallees, banksias, coral heath and many flowering peas. Eucalypt and blackbutt forests give way to subtropical rainforests along the Hacking River, where turpentines, coachwoods, sassafras and cabbage tree palms are important species.

Walking Tracks

Numerous well-maintained and signposted walking tracks cross the park, from easy nature trails to long bushwalks. The coastal tracts meander along the unspoilt coastline, along cliffs, beaches, though forests and over heathlands. Inland trails pass waterfalls, freshwater pools and creeks and provide excellent views over the park and coast. Aboriginal rock carvings and shell middens can be seen on some of the walks.

Other Activities

Boating at Port Hacking and on the river at Audley is very popular, canoes and rowboats can be hired at Audley. There are many beautiful picnic spots and good beaches for swimming and surfing. Many people fish from the rocks and beaches.

The park offers good opportunities for birdwatching and wildflower study.

Climate

Like Sydney, the park has warm summers and cool winters, with the highest rainfall in the summer months. Visit all year round.

Bookings and Fees

Bookings are accepted and fees charged for the camping area at Bonnie Vale. Park entry fees are payable.

Information and Bookings

The Superintendent
Royal National Park
P.O. Box 44
Sutherland
New South Wales 2232
Tel. (02) 521 2230

Sturt
National Park

In the far north-west corner of New South Wales, where red sand dunes stretch to the horizon, and flat-topped rocky mesas tower over parched gibber plains, lies the remote yet fascinating Sturt National Park. Here, in the true outback Australia, is an isolated area set aside to preserve the unique flora and fauna of our arid inland regions.

Facilities

There are car and bus camping sites with picnic areas at Dead Horse Gully (1 km north of Tibooburra), Mt Wood, Olive Downs and Fort Grey, all with toilets, barbecues and water. A visitor centre is located at Tibooburra.

Although bush camping may be permitted, it is not recommended because of the isolated nature of the area and the harsh climatic conditions. Approval must be obtained from the officer-in-charge.

Accommodation, fuel and stores are available at Tibooburra and Milparinka.

Location and Access

The park is 1500 km north-west of Sydney, 330 km north of Broken Hill via the Silver City Highway, and 430 km west of Bourke via Wanaaring. These roads and those within the park are unsealed and may become impassable after heavy rain.

All visitors should carry extra food, water, fuel and vehicle spares, and should remain with the vehicle in the event of a breakdown.

Landscape

This vast park covers 307 507 ha of semi-desert country. Floodplains dotted with occasional trees in the east of the park give way to small rocky gorges, dry creek beds lined with river red gums and 'jump-up' country around Olive Downs, where flat-topped mesas rise up to 150 m above the plain. Further west, red-soil gibber plains are replaced by the sandhills of the Strzelecki Desert.

Flora and Fauna

Red and grey kangaroos, euros and emus are frequently seen. Among the reptiles are bearded dragons, shingleback and gidgee lizards, sand goannas, western brown and common brown snakes.

The magnificent Sturt's desert pea thrives after rain in Sturt National Park.

Lake Pinaroo fills with water after heavy rain, and is an ideal birdwatching area. Birds of prey include the wedge-tailed eagle and numerous other raptor species. There are cockatiels, galahs, little corellas, parrots, pigeons, corvids, honeyeaters, babblers, inland dotterels and the pratincole, also called the road runner.

The common tree species include the river red gum, coolabah, bloodwood, gidgee, mulga, whitewood and beefwood. The low shrub layer is dominated by saltbush, bluebush, copperburr and Mitchell grass. After good rains the desert comes alive with rapidly growing herbs such as the black and scarlet Sturt desert pea, purple parakeelia and carpets of white and yellow daisies.

Walking Tracks

There are well-marked walking tracks through the 'jump-up' area varying from one to three hours, and an easy walk through the granite outcrop near Dead Horse Gully. A walking trail leads to Mt Wood summit from Mt Wood Road for an excellent overview of the area. The scenery is fascinating and there are many opportunities to study the wildlife of the area. Drinking water should be carried in the warmer months.

Other Activities

The park has a pastoral museum at Mt Wood, various relics of the early cattle and sheep rearing days, and recreated gold-mining site at Golden Gully. Visitors can drive along some of the 300 km of roads where there are many excellent opportunities for photography and nature study.

Climate

The area has an average annual rainfall of only 225 mm, with very hot summers. Visit in winter and spring.

Bookings and Fees

No bookings are accepted, and a camping fee is payable at Dead Horse Gully only.

Information

National Parks and Wildlife Service
Tibooburra District Office
Briscoe Street
Tibooburra
New South Wales 2880
Tel. (080) 91 3308

Volcanic plugs on the Grand High Tops overlook the valleys of the Warrumbungle National Park.

Warrumbungle

National Park

West of the Great Divide, where the endless dry plains of central Australia begin, the Warrumbungles rise majestically in a series of barren spires, forested ridges, tabletops and deep, plunging gorges. Encompassing an area of 19 651 ha, the Warrumbungle National Park is a sanctuary for a rich and varied mammal and bird population, a place of breathtaking views and spectacular springtime wildflower displays.

Facilities

There are five campsites with vehicular access: Camp Blackman, Camp Wambelong, Camp Elongery, Camp Pincham (gear must be carried from nearby carpark and Camp Burbie (four-wheel-drive only with permission). Camp Blackman has 32 powered caravan sites, septic toilets, hot showers and

The green frog, often seen in the Warrumbungles.

basic laundry facilities, the others have pit toilets and tank water. Camp Wambelong and Camp Elongery are used by large groups.

There is a visitor centre and kiosk near Camp Elongery. Balor Hut below the Breadknife can accommodate six people, and the Woolshed can be booked by large groups.

Backpackers can camp at designated places on the walking trails. Water must be carried and permits obtained from the ranger.

Accommodation is available in Coonabarabran.

The looming spire of the Breadknife (80 m) in the Warrumbungle National Park.

Location and Access

The park is 491 km north-west of Sydney. It can be reached from the east via Coonabarabran (37 km), from the west via Coonamble, and from the south via Gilgandra, Tooraweenah and Gumin. Visitors are requested to register with the ranger at the visitor centre near Camp Elongery. Roads within the park are almost all sealed.

Landscape

The park is part of the Warrumbungle Mountain range, the remnant of a large volcanic centre active some 13 million years ago. Rocky spires, domes, sandstone plateaux, bluffs, and a spectacular 90 m high wall, the Breadknife, dominate the landscape. Years of erosion have left hard volcanic plugs which now tower over the wooded countryside, and deep river valleys have cut through softer rocks revealing ancient sandstone beds.

Flora and Fauna

Grey kangaroos, wallaroos, red-necked wallabies, possums, anteaters and gliders are common. Koalas can sometimes be found in woodlands.

Emus, wedge-tailed eagles, parrots and cockatoos are frequently seen.

There are numerous reptiles including the bearded dragon, lace monitor, skinks, blue-tongued lizard, geckos, green tree frogs and some snakes.

The rich flora of the area includes eucalypts such as the white gum, white box and stringybarks; blackboys, mallee and wattles are common; water conserving plants typical of the western plains are found in drier areas. Numerous wildflowers bloom in spring, notably purple *Hovea lanceolata,* flannel flowers, boronias, scented sun orchids and pimelias.

Overlooking the camping ground at Warrumbungle National Park with the barren Warrumbungle peaks rising majestically in the background.

Grey kangaroos on the heath plateau in the Warrumbungles.

Walking Tracks

There are 11 guided walks from 1 km to 16 km, taking from 30 mins to 6 hrs. They include an easy nature trail and many spectacular walks through valleys and across mountain passes, all with splendid views. The trails are marked and maintained, but water must be carried.

Other Activities

The park is one of the best places in Australia for rock climbing, which is allowed anywhere except on the Breadknife and Chalker's Mountain. There are many easy, difficult and classical climbs, although permits are required.

The extreme clarity of the air makes the park a good spot for astronomical observation and photography. Siding Spring Observatory is close by.

There are many opportunities for birdwatching and wildflower study.

Climate

The park has hot summers and cold winters, with a low annual rainfall. Visit all year round.

Bookings and Fees

Bookings are required for Balor Hut, the Woolshed and by groups camping at Camps Wambelong and Elongery. No advance bookings are taken for camping or powered caravan sites. Park and camping fees are payable.

Information and Bookings

Ranger-in-charge
Warrumbungle National Park
P.O. Box 39
Coonabarabran
New South Wales 2857
Tel. (068) 25 4364

Queensland

Queenland's National Parks offer endless stretches of shimmering sands and pounding surf; a coast of coral reef and great sandy beach and dune systems. Inland are rugged ranges, high waterfalls, dense rainforests and fascinating tropical wildlife.

The north-east rainforests are visually overpowering; great columns of tree soar upwards to a canopy of dense foliage that often blocks out the sky. The north-east coastline is equally magnificent; here is a tremendous variety of cliffs and long golden beaches, forests right to the water's edge, high sand dunes, clear water and colourful corals and dark mangrove swamps. The Great Barrier Reef is the exciting culmination of this magnificent coastal scenery—the most massive structure erected by living creatures on the face of this planet, the reef extends some 2000 km.

The Queensland coast changes as it moves south, becoming dominated by long, golden sandy beaches rather than hills and forests, with extensive heathlands and swamps behind the beaches. The islands are huge sand dunes rather than the drowned ranges of coral cays in the north. Islands off the coast protected by National Park status include Hinchinbrook, Magnetic, Dunk, Heron, Lindeman, and the world's largest sand island, Fraser.

Within the northern coastal area there are approximately 40 mountain parks, all excelling in rugged river gorges, spectacular waterfalls, placid lakes, lush rainforests and a wide variety of wildlife.

Within the central coastal region the National Parks are mostly coastal islands but also embrace two outstanding parks in the hinterland. Eungella National Park, one of Queensland's wildest and most majestic parks, is an oasis of rainforest capped in cloud, overlooking a coastline of mountainous islands.

In the heart of the Great Dividing Range is Carnarvon National Park, one of the most breathtakingly scenic reserves in Australia.

Long sweeps of beaches and sandy islands, highland forests, streams and fine wildlife habitats feature in the southern coastal region of Queensland. The overwhelming number of parks in this area include Tamborine Mountain, famed for its waterfalls; Lamington, rich in rainforests, ferns and orchids; and Noosa, with a wide variety of vegetation and coastal scenery. Close to the New South Wales' border is Girraween National Park, the place of flowers, a photographer's paradise. This park offers the best floral displays in the state.

Blackdown Tableland
National Park

This undulating plateau has a cooler, moister environment than the surrounding land and offers spectacular sandstone scenery including cliffs and pinnacles, gorges and waterfalls. It is a place to relax, enjoy, explore and cool off in the several streams that flow throughout the year. The Tableland remained inaccessible to all but stockmen and occasional adventurous visitors until 1974 when a Forestry Department road was built to enable the harvesting of extensive eucalypt stands.

Facilities
Visitors should be self-sufficient; fuel and food and a caravan park are available at Dingo. There is a camping area with toilets near South Mimosa Creek, and Horseshoe Lookout and Officers Pocket have tables, seats and fireplaces for day visitors. Several streams can be relied on for water at all times and visitors are advised to bring containers to carry water from North Mimosa Creek, about 3 km by road from the campsite. The water should be boiled for drinking.

A permit to cross areas of state forest and a permit to camp are required, both are available from the ranger.

Tree ferns adorn the gorges in Blackdown National Park.

Location and Access
Blackdown Tableland is located 160 km west of Rockhampton; 55 km south-east of Blackwater. Car access is from the Capricorn Highway, 11 km west of Dingo. This is a gravel forestry road, unsuitable for caravans.

Access may be restricted in wet weather or in high fire-risk periods. A four-wheel-drive vehicle is necessary for only one road within the park.

Landscape
Blackdown Tableland, a two-part park of nearly 24 000 ha, occupies an undulating plateau of sandstone, 800–950 m above sea level. The tableland stands prominently above the plains of the lower Bowen basin. This convergence of the Expedition, Shotover and Dawson Ranges is the northern extension of the central Queensland sandstone region. The plateau is cut by Mimosa Creek and its various tributaries into rugged gorges.

Flora and Fauna
Tall, dense sclerophyll forest covers the plateau and waterfalls drop into pools fringed with cabbage tree palm, treeferns and various other ferns. Lily pillies grow along some creeks and the cliffs and sandstone outcrops support elkhorns and orchids. Heathlands occur in shallow soil. Several species of plants found here, including a wattle, a callistemon, macrozamia and stringybark, occur nowhere else.

Leaching of most of the soil phosphates results in a bone deficiency in foraging animals so there are few large mammals like kangaroos or wallabies found in the park. However, being isolated by high cliffs, the tableland has many species of animals that are found nowhere else. These include a lizard, a trapdoor spider and a Christmas beetle. The forests also support a rich and varied bird life.

Walking Tracks
A variety of walking tracks cover about 20 km. Climbing is not advised as the edges of the sandstone cliffs may be soft and crumble. The most popular areas to visit are Rainbow Falls, the western lookouts and Stony Creek Falls. Visitors are advised to check walking plans with the ranger.

Other Activities
Exploring, nature appreciation, photography and swimming are popular activities within the park.

Climate
Spring and autumn are the best seasons to visit. Major rainfall occurs from December to March when the area is subject to dense fogs.

Bookings and Fees
Fees are payable for camping and the use of facilities including guided tours and other activities. Bookings are required during the holiday periods.

Information and Bookings
The Ranger
Blackdown Tableland National Park
C/o Dingo
Queensland 4702
Tel. (079) 86 1964

Queensland National Parks and Wildlife Service
Quay Street
Rockhampton
Queensland 4700
Tel. (079) 27 6511

Bunya Mountains National Park preserves the largest remaining natural stand of bunya pines in Australia.

Bunya Mountains
National Park

Overlooking the Darling Downs and the extensive eastern plateau, the Bunya Mountains National Park is just a small part of the Great Dividing Range, yet it preserved from the timberfeller's axe one of the last remaining forests of bunya pines in Queensland. An important rendezvous for Aboriginal tribes from a wide area who used to come to feast and celebrate the bounteous crop of delicious bunya nuts, the park today has lost its original inhabitants, but remains a place of beautiful rainforest, spectacular waterfalls and colourful bird life.

Facilities
The main car camping area at Dandabah near the ranger station has fireplaces, toilets, showers, a kiosk and picnic area. Burton's Well campsite has fireplaces, toilets, water and picnic tables. Westcott campsite offers fireplaces, toilets and water. Other picnic areas have been established with fireplaces and tables. Bush camping is permitted, but must be more than 3 km from roads and facilities.

Accommodation is also available in privately owned motel units, a guest house and cabins near the park entrance.

Location and Access
The park is some 250 km north-west of Brisbane via the Warrego Highway through Toowoomba, turn off at Jondaryan, and follow the signs to Bunya Mountains travelling through Quinalow. From Dalby travel via Kaimkillenbun and Yamison. From Kingaroy access is via Kumbia, and from Nanango via Maindenwell. All approaches are steep and towing caravans up the mountain is not recommended. Access roads within the park and from Kingaroy are sealed.

Landscape
The park occupies 11 700 ha, straddling a ridge rising to 1135 m at Mt Kiangarow, and forming part of the Great Dividing Range. A number of creeks originate in the park, descending in cascades and waterfalls over cliffs. There are several forms of rainforests, woodlands, open forests and grasslands in the park.

Flora and Fauna
Red-necked wallabies feed on open grasslands, while swamp wallabies and red-necked pademelons are found in the rainforests. The Bunya Mountains ringtail possum is only found in this area and carpet snakes are often seen on the trails. King parrots and crimson rosellas are a great attraction in picnic areas with their brilliant plumage.

Towering bunya pines dominate the high rainforests, being replaced by hoop pines lower down. Epiphytic ferns, orchids, mosses and lichens inhabit wet rainforests. Australian bottletrees and brigalow-belah are found in dry rainforests. There are grassy eucalypt forests with Queensland blue gums, narrow-leaved stringybarks, yellow and white box and grass trees.

Walking Tracks
An extensive system of graded walking trails has been constructed, including a walk for disabled people at nearby Russell Park. They vary from a 500 m nature walk to a 10 km hike to Big Fall Lookout. The varied track system has many rewards for keen bushwalkers.

King parrots are a popular attraction in Bunya Mountains National Park.

Delicate ephiphytic ferns grow in the wet rainforest areas of Bunya Mountains.

Climate
The park has mild, wet summers with cool nights, and cold winters when temperatures fall to freezing point at night. Visit all year round.

Bookings and Fees
Camping permits must be obtained from the ranger, bookings are accepted up to 12 weeks in advance. Fees are payable for camping and the use of facilities including guided tours and other activities.

Information and Bookings
Ranger-in-charge
Bunya Mountains National Park
M.S. 501
Dalby
Queensland 4405
Tel. (074) 68 3127

Cape Tribulation
National Park

Cloud-shrouded mountain tops and rainforests sweep down from the heights of the Thornton Range to a stretch of sandy beaches and rocky headlands where Cape Tribulation National Park meets the Great Barrier Reef. Walkers can explore a wonderful diversity of plants and wildlife in a largely undeveloped area.

Facilities
Noah Beach, 8 km off Cape Tribulation, has a camping area where toilets and water are provided. A permit, obtainable from the ranger, is required. There is a public camping ground at Thorntons Beach and private camping areas at Myall Creek and Cape Tribulation where a youth hostel and holiday units are also sited.

Sun filters through the rainforest canopy in Cape Tribulation.

The colourful noisy pitta found at Cape Tribulation National Park.

A picnic area at Cape Tribulation has water, and toilets, and a nearby shop provides groceries and fuel. There is also a kiosk at Thorntons Beach and fuel and light refreshments available at a teahouse north of the Daintree River.

Location and Access

The park is situated 140 km north of Cairns. Car access is along the Cook Highway via Mossman to the Daintree River ferry. From the ferry the road is unsealed and four-wheel-drive vehicles are essential after heavy rain, especially from December to April. During very wet periods the road is sometimes closed completely.

Landscape

The 16 965 ha of Cape Tribulation National Park extend along the coast from the Daintree to the Bloomfield River and up into the McDowall Range in the west to Thornton Peak (1375 m), the park's highest summit. Many creeks traverse the park, running through boulder-strewn gullies and dense rainforest.

Flora and Fauna

Within the park boundaries vegetation varies from rainforest to eucalypt forest, swamp communities and mangroves. The rainforest contains a number of rare, primitive flowering plants including a species, *Idiospermum australiense,* the only representative of its family. The trunks of the rainforest canopy are hung with a variety of ferns, vines and orchids while the forest floor is deep with decomposing logs and fallen leaves. Bennett's tree kangaroo is found only in the rainforests of this region. Bird life includes the Wompoo pigeon, colourful noisy pitta and the white-tailed kingfisher.

Walking Tracks

There are no marked tracks in the park but the coastline and creeks provide relatively easy walking.

Other Activities

Guided bushwalks are offered by private companies as are day trips to the Reef. Near the Daintree River, ferry boat ramps allow the launching of small craft.

Crocodiles are present in the creeks and marine stringers are found in coastal waters from October to April.

Climate

Winter is the best time to visit the park as summer is hot and very wet and swimming is unsafe at this time.

Bookings and Fees

Fees are payable for camping and the use of facilities including guided tours and other activities. Permits to camp are issued on site only. Fees are also payable for private camping grounds and other accommodation.

Information and Bookings

Ranger
Cape Tribulation National Park
P.M.B. 10, P.S. 2041
Mossman
Queensland 4873

Carnarvon

National Park (Carnarvon Gorge Section)

Carved from the white sandstone of the Great Dividing Range, more than 400 km from the coast of central Queensland, the mighty Carnarvon Gorge appears as a cool green oasis, a lost fertile valley in the dry highland plateau. The heart of Carnarvon National Park, the Gorge is the grand master of a maze of narrow, twisting ravines etched from the rock by the Carnarvon Creek and its tributaries. For 30 km the creek twists through towering cliffs 400 m wide and 200 m high, a repository of Aboriginal art and sanctuary for rich floral and faunal communities.

Facilities

A camping area is located by the creek near the park entrance, with toilets, cold showers and a visitor centre. Firewood is not available, and visitors should bring their own stoves. Advance booking is essential. There is a walk-in camping area with toilets at Big Bend, 500 m upstream of Cathedral Cave.

Cabbage palms cluster in the damp gorges of Carnarvon National Park.

Bush camping is permitted, but must be at least 10 km upstream from the main camping area. Always check with the ranger before preparing to bush camp.

Accommodation is also available at the private tourist lodge near the park entrance, where basic provisions can be purchased.

Location and Access

The park is 765 km north-west of Brisbane via Roma, Injune and Wyseby. The road from Roma is sealed for 170 km and unsealed for 75 km. From Rollestone to the north of the park the road is sealed for 20 km and the remainder of the 100 km is unsealed. Both roads become impassable after heavy rain, and visitors should take several days' extra food. The road from Rolleston is better for caravans.

Landscape

The park occupies 28 500 ha of the Consuelo Tableland and the Great Dividing Range, rising to 1232 m on the Consuelo Tableland. Cliffs, canyons and gorges are the most prominent features, with Carnarvon Gorge breaking from the range between huge white walls nearly 2 km apart at its entrance. The high, dry, plateau country of rugged hills and gentle slopes supports grassy open eucalypt forests, while the gorge floor is green, fertile and richly vegetated.

Flora and Fauna

Grey kangaroos brush-tailed rock wallabies, whiptail wallabies, brush-tailed possums and greater gliders live in the forests. Platypuses, perch, eels, catfish, freshwater snakes, the common tree snake and eastern water dragon may be seen swimming in the creek, while green tree frogs are often found on the banks.

Among the many birds are dusky moorhens, azure kingfishers, herons and pied cormorant found by the creek, with yellow-tufted honeyeaters, fairy wrens, currawongs and apostle birds nearby.

A friendly greeting from an inhabitant of Carnarvon National Park.

Left: *Climbing the Amphitheatre on the south side of Carnarvon Gorge.*

Aboriginal rock paintings in the Art Gallery at Carnarvon National Park.

Near the creek grow she-oaks, weeping red bottlebrushes, the native cherry, ferns and mosses. Higher up are cabbage tree palms, cycads, treeferns, forest red gums, native hibiscus, figs and bracken ferns. Further from the creek are spotted gums, grey gums and grass trees. Wards Canyon is the only place in the park where the king fern, *Angiopteris evecta*, grows.

Walking Tracks

Walking tracks follow the main gorge and its tributaries, leading to Aboriginal rock paintings and carvings, caves and lookouts. The main gorge and rugged tablelands offer extensive bushwalking possibilities.

Other Activities

The area is a delight for photographers and rich in Aboriginal art. In the outer gorge section there is a waterhole suitable for swimming.

Climate

Summers are warm to hot, winters cold to warm. Most rain falls from December to March. Visit between April and November.

Bookings and Fees

Campsites must be booked in advance and permits obtained from the ranger. Fees are payable for camping and the use of facilities including guided tours and other activities.

Information and Bookings

The Park Ranger
Carnarvon National Park
Via Rolleston
Queensland 4702
Tel. (079) 84 4505

Chillagoe-Mungana Caves
National Park

Due west of Cairns the lush tropical rainforests of the Atherton Tableland give way to a vast, dry, undulating plain dotted with spectacular, vertical-walled limestone towers. These huge, dark grey edifices with their projecting ridges and great, jagged fissures resemble the broken battlements of ancient castles. Below them an extraordinary undergound cave system has formed, extending for many kilometres, decorated with interesting flowstone formations. The most dramatic caves have been protected in the Chillagoe-Mungana Caves National Park which actually comprises nine individual National Parks.

Facilities

There are car camping sites near Royal Arch Cave in the Chillagoe Caves area with picnic tables, fireplaces and toilets. Visitors must carry their own water.

Accommodation is available in Chillagoe, where there are motels, hotels and caravan parks.

Delicately feathered grass whistling ducks seen at Chillagoe-Mungana Caves National Park.

Location and Access

The park is 1985 km north-west of Brisbane and 215 km west of Cairns via Mareeba and Dimbulah. The road is mostly unsealed from Dimbulah and may be impassable to ordinary vehicles after heavy rain. One train a week goes from Cairns to Chillagoe as well as a bus service, and charter flights can be arranged.

Landscape

The nine parks which comprise Chillagoe-Mungana Caves National Park are Royal Arch Cave, Tower of London and Donna Caves, Jubilee and Piano Caves, Cathedral Cave, Geck and Spring Caves, Eclipse Cave, Ryan Imperial Cave, Markham Cave and Royal Archway Cave. These cover 1876 ha of an eroded limestone belt with jagged outcrops rising up to 70 m above the plains. Severe erosion by tropical rains has formed an extensive cave system inside and below the outcrops. Richly coloured cave coral stalactites, stalagmites, shawls and pendants are found in huge caverns and labyrinthine tunnels.

Flora and Fauna

The stable cave environments are ideal for many bat species including the bent-wing bat, the little brown bat, the sheath-tailed bat and the horseshoe bat. The children's python and brown tree snake prey on the bats, and the caves are one of the few known nesting sites of the grey swiftlet, an unusual bird species with an echo-location system.

The preserved bones of dingos, wallaroos, brush-tailed rock wallabies, red-legged pademelons, marsupial mice, short-nosed bandicoots, northern quolls, sugar gliders, phascogales and blue-tongued lizards are found in the caves.

Above ground is a scrubby type of vegetation with many deciduous trees, notably the bats-wing coral tree and the narrow-leafed bottle tree. Native bauhinias and the helicopter tree (a species of Gyrocarpus) are common, with fig trees on the outcrops and some ghost gums on the plains.

Walking Tracks

Several major caves of Chillagoe can be explored. The ranger conducts guided tours through three caves, and others can be seen independently. Caves must not be entered without the ranger's permission.

Royal Arch Cave is one of the largest cave systems and is an easy ramble, carrying torches, through 1.5 km of passages and chambers. Donna Cave is lit, and is perhaps the most beautiful. Trezkinn Cave is noted for its magnificent chandelier.

Self-guided tours can be made through The Archways with its giant grike system and wildlife, Pompeii Cave, a huge rock cleft, and a number of other caves. A good torch and strong shoes are essential.

Walking tracks above ground lead to interesting formations and lookouts.

Jagged limestone ridges at Chillagoe-Mungana Caves National Park.

Impressive stalactite and stalagmite formations in Donna Cave in the Chillagoe-Mungana Caves National Park.

Other Activities

Disused copper smelters and mines are of interest to mineral collectors. A few small galleries of Aboriginal paintings occur in the area. Swimming is possible in the creeks at designated areas.

Climate

There is a wet season from January until April, when most of the 800 mm of rain falls. October and November are very hot and dry. May to August is the most pleasant time to visit.

Bookings and Fees

Bookings may be made for the campsites up to three months in advance. Fees are payable for camping and the use of facilities including guided tours and other activities.

Information and Bookings

The Ranger
Chillagoe-Mungana Caves National Park
P.O. Box 38
Chillagoe
Queensland 4781
Tel. (070) 94 7163

Conway, Cumberland and Whitsunday Islands

National Parks

Along the Whitsunday Passage there are more than 90 islands which are the peaks of a mountain range once joined to the mainland. They extend in a 160 km arc from Hayman Island in the north to Penrith Island in the south of the Cumberland group. Apart from Hayman, Dent and Hamilton Islands, together with some other very small islands, the whole group is now a part of the system of National Parks in the Whitsunday and Cumberland Islands which all exist within the Great Barrier Reef Marine Park.

Facilities

Owing to the number of islands an exhaustive coverage is difficult but the following will provide a guide.

This craggy promontory with surrounding crystal clear waters forms part of Whitsunday Island National Park.

Cruising is the most popular pastime in the Whitsundays.

Bacchus

Panoramic view of the Whitsunday Islands.

Gloucester Island is accessible by private boat from Bowen or Dingo Beach. Bona Bay on the south-western side is an excellent anchorage where pit toilets and picnic tables are provided but no fresh water is available. Other western beaches have no facilities but camping by permit is allowed and there are good anchorages.

Middle Island is accessible by private boat from Bowen. The sandy beach on the western side is suitable for camping although there are no facilities or fresh water. Flatback Turtle, *Chelonig depressa,* nest here.

On Whitsunday Island the more popular bays include the Cid Harbour area, Whitehaven Beach, Chance Bay and Turtle Bay. Cid Harbour provides camping areas with facilities and an excellent anchorage for all winds. Whitehaven Beach's facilities include pit toilets and picnic tables in the camping area.

Haselwood, Lupton and Edward Islands are accessible by private boat.

Hook Island can be reached by tourist and private boats. Nara Inlet is a popular safe anchorage.

Tourist resorts on National Park islands are at South Molle, Long and Lindeman Islands. Camping is not permitted on National Park or tourist resort islands. Other resorts in the group are Hayman Island, Daydream Island and Hamilton Island. Hook Island has a private camping area and an underwater observatory as well as an Aboriginal art site.

Conway has a 16-site camping area opposite the park headquarters. Toilets, fireplaces and tables are provided. A time limit of four nights is enforced in busy periods. At Swamp Bay there is a small bush camping area.

Shute Harbour, the gateway to the Whitsunday and Cumberland Islands, is a safe anchorage for all forms of boats.

Location and Access

The islands are situated between 50 and 120 km north of Mackay. They are accessible by boat from Cannonvale, Airlie Beach and Shute Harbour and by air from Mackay, Proserpine and Hamilton Island. Light plane airstrips exist on Lindeman and Brampton Islands and at Shute Harbour. Float planes also operate out of Shute Harbour.

Conway National Park, a large mountainous rainforest area on the mainland, is reached via the Bruce Highway from Proserpine, turning off to Airlie Beach. The camping ground, which includes a day-use area, is 9 km beyond the township on the Shute Harbour Road.

Landscape

Most of the islands have rough and hilly terrain with plunging cliffs on the headlands and craggy peaks and domes of bare rock on the hillsides. Stream flow varies from rushing torrents to isolated waterholes in dry periods. Hook and Whitsunday

Islands have fiord-like inlets around which the hills rise steeply to over 300 m. Many smaller islands have bays lined with sandy beaches and fringed by coral reefs which offer good anchorage for launches and yachts. Proper anchoring techniques must be used and it is forbidden to anchor on coral. A significant tidal range of up to 6 m and tidal surges up to 8 km/hour can make boating hazardous near headlands and passages. Diving should be done on the turn of the tide.

Flora and Fauna

Throughout the islands vegetation is generally dense comprising grasslands, acacia woodlands, open eucalypt forests and rainforests, eucalypts with emergent hoop pines.

Mammal and reptile species on the islands are few in comparison with the adjacent mainland. Nocturnal brushtail possums, sugar gliders and bandicoots are found at Conway while on the islands the black flying fox, lace monitor and fawn-footed melomy are most common. A wide range of bird life includes noisy pittas, orange-footed scrub fowls, ospreys, white-breasted sea eagles and brahminy kites.

Walking Tracks

South Molle, Lindeman, Long and Whitsunday Islands as well as Conway National Park have marked walking trails. Otherwise, bushwalking is available on most of the islands, ranging from easy to very rough conditions.

Other Activities

Swimming, fishing, boating and snorkelling in approved zones are the most popular activities on the islands. Hook Island contains accessible Aboriginal cave drawings with a developed boardwalk within the cave, while Dent Island has a lighthouse and coral art display. During summer marine stingers may be present and swimming is dangerous particularly near the mainland. Most island beaches are free from stingers. The ranger can provide detailed information.

Climate

Heavy rain falls from December to April when temperatures are very hot and cyclonic storms sometimes occur. The most pleasant times to visit are in autumn and spring.

Bookings and Fees

Campers are required to book no more than 6 weeks in advance for their permit. Information regarding privately managed resorts can be obtained from the Queensland Government Travel Centre. Fees are payable for camping and the use of facilities including guided tours and other activities.

Information and Bookings

Ranger
Conway National Park
P.O. Box 332
Airlie Beach
Queensland 4802
Tel. (079) 46 9430

Right: *Sunset over the Barrier Reef island parks as seen from Cardwell.*

A pandanus palm overlooks sheltered rock pools in the Whitsunday Passage.

Left: *Vigorous mangrove forests cloak the foreshores along the north Queensland coastline.*

Hoop pines surround a quiet bay by on Magnetic Island National Park.

Cooloola
National Park

Stretching from the salty mangrove-fringed shores of Lake Cootharaba north of Noosa Heads through areas of open heathland that form a mosaic of wildflowers in spring, Cooloola National Park reaches north-east along the rainbow-coloured sand cliffs of the coastline to the peaceful waterways of Tin Can Bay.

Facilities
The administration of Cooloola is divided into two sections with rangers stationed in the north at Freshwater camping area and in Southern Cooloola at the Sir Thomas Hiley Visitor Centre near Kinaba Island.

Northern Cooloola's main camping area is located near the beach, 1 km from Freshwater Lake where barbecues, tables, picnic areas, water and toilets are provided. West of Double Island Point there is a camping area with septic toilets and along the beach camping sites without facilities are available. Camping is allowed elsewhere in the park but a permit must be obtained from the ranger. Other picnic areas with barbecues and tables can be found along the Freshwater Road at Blackbutt, Bymiem and Quandong.

South Cooloola has privately owned camping facilities with showers, toilets, barbecues and a shop at Elanda Point and Lake Cooribah Holiday Park. North of the visitor centre, Harry Spring's Hut is a small camping zone with toilets, barbecues and picnic tables and is accessible by boat or four-wheel-drive vehicle. Along the upper Noosa River there are a number of other small sites.

There are no shops in the park but accommodation and supplies are available at the nearby townships of Noosa, Tewantin, Boreen Point and Rainbow Beach.

Location and Access
The park is situated about 30 km west of Gympie. In the north the Rainbow Beach township is reached by bitumen road turning off the Bruce Highway at Gympie. Access to Double

Island Point and Freshwater is via the beach in a four-wheel-drive vehicle. Branching off the Rainbow Beach Road, 4 km before the township, a signposted sand road leads to camping and picnic areas and walking tracks. Conventional vehicle access is sometimes possible but a four-wheel-drive is recommended.

From Tewantin across the Noosa River a vehicular ferry gives four-wheel-drive access to the beach's southern end. In the west, Cooloola Way is a gravel road which links Kin Kin with the Rainbow Beach Road. Access to the Noosa River at Harry Spring's Hut is possible via a rough track branching off the Cooloola Way. Again four-wheel-drive vehicles are preferable.

Commercial tours leaving from Noosa, Tewantin and Brisbane reach various parts of Cooloola by boat, canoe, vehicle and on foot. Canoes may be hired at Boreen Point and Elanda Point and small power boats are available for hire at Boreen Point, Elanda Point and Noosaville.

Landscape

Cooloola National Park's 40 900 ha include areas of mangrove-lined waterways, tranquil lakes and rainforests. The coastal region is notable for its high dune formations, wind driven sandblows and multi-coloured sandcliffs.

Inland Cooloola is a wilderness of sandy heathlands, swamps and forested hills surrounding the upper Noosa River, its tributaries and Lake Cootharaba.

The graceful brolga may sometimes be seen around Cooloola National Park.

Left: *Four-wheel-drive vehicles are a popular means of exploring the coloured sands at Cooloola.*

Flora and Fauna

The waterways are fringed with sedges, mangroves and casuarinas while the swampy heathlands, studded with grass trees, hibiscus, melastoma and other wildflowers reach into densely wooded dune ridges of banksia, scribbly gum and blackbutt forests. Rainforest areas comprise a variety of palms, ferns and vine forests.

Animals include nocturnal mammals such as possums, bandicoots, native rats and marsupial mice as well as some potoroos and pademelons. Rainbow lorikeets, numerous sea birds, mangrove herons, waterfowl, parrots and bowerbirds comprise the prolific bird life of the park.

Walking Tracks

A range of walking trails of various lengths traverse the park. Some are self-guided such as the circuit track located near the visitor centre which includes a section of boardwalk through the mangroves. Brochures detailing the trails may be obtained from the rangers. Walkers who plan to leave the marked tracks or take extended walks must inform the ranger.

Other Activities

Surf, lake and river swimming are much enjoyed at Cooloola particularly in the summer. Canoeing enables leisurely travel on the Noosa River and motor boats can be used provided the 4-knot speed limit upstream of the visitor centre and defined landing points is adhered to for the preservation of the environment.

Climate

During summer from December to February temperatures average above 30°C in the day time and rainfall and humidity are high. Autumn, winter and spring enjoy a very pleasant climate with warm sunny days and cooler nights.

Multi-coloured sand cliffs border the extensive coastal stretch at Cooloola.

Bookings and Fees

Fees are payable for camping and the use of facilities including guided tours and other activities. Campers in all National Park grounds must possess a permit. Bookings are required and essential for holiday periods. These can be made 6–12 weeks in advance through the Rainbow Beach office for the northern section of the park and the Elanda office for Harry Spring's Hut and upper Noosa River campsites in the south.

Information and Bookings

Ranger
Sir Thomas Hiley Visitor Centre
Queensland National Parks and Wildlife Service
Elanda
Via Tewantin
Queensland 4565
Tel. (071) 85 3245 or (071) 49 7364

National Parks and Wildlife Service
P.O. Box 30
Rainbow Beach
Queensland 4570
Tel. (071) 86 3160

The white-faced heron thrives on the abundant sea life in the waters of Cooloola National Park.

Eungella
National Park

Inland of Mackay on the central Queensland coast, rugged mountains clothed in luxuriant rainforests rise from green expanses of sugar cane covering the foothills and coastal plains. Eungella National Park straddles the Clark Range, extending for 35 km along misty, towering mountain peaks and deep gorges. It is perhaps Queensland's wildest and most majestic National Park, a place to enjoy the cool mountain air, walk below the cathedral-like vaults of the rainforest canopy and even observe platypuses paddling in the crystal pools.

Facilities

A caravan and car camping area is located at Broken River, 6 km south of Eungella, where there are toilets, fireplaces, cold showers, picnic tables, a kiosk and ranger station. A tent camping area is located 500 m from the existing Broken River camp area. An information centre is located in the picnic area. The camping area at Finch Hatton Gorge has toilets, fireplaces and picnic tables, but the road is not normally suitable for conventional vehicles. Bush camping is permitted, but must be more than 3 km from roads and facilities and the ranger must be notified beforehand.

Accommodation is available at Eungella.

Rainforest ferns thrive on the filtered sunlight beneath the high rainforest canopy in Eungella National Park.

Location and Access

The park is 1110 km north-west of Brisbane and 83 km west of Mackay along the sealed Eungella road, turn off at Eungella along the sealed Broken River Road to the park. Large caravans will have difficulty negotiating the Eungella road, and care must be exercised on the dirt road after rain.

Landscape

The park occupies 50 800 ha of extremely rugged mountain terrain rising to 1280 m at Mt Dalrymple, and descending in steep rainforest slopes and deep gorges cut by fast-flowing streams to the flat, fertile Pioneer Valley. The northern section of the park is a rugged, trackless wilderness area. The southern part around Broken River has been sufficiently developed to allow visitors access to rainforests, rockpools and viewpoints; and the Finch Hatton Gorge area provides access to creeks and waterfalls.

Views from the rugged rainforested peaks of Clark Range looking over the canefields and coastline near Mackay.

The fast-flowing tract of Broken River plummets in a series of rapids and cascades in Eungella National Park.

Flora and Fauna

Red-legged pademelons, fawn-footed melomys, long-nosed bandicoots, the brown marsupial mouse, possums, gliders and dingos inhabit the park. Platypuses are often seen in the pool near the Broken River campsite. There are more than 100 bird species including many species of brilliantly coloured rainforest pigeons, the regent bowerbird, brown thornbill, white-browed scrub wrens and pied currawongs.

The rainforests are rich with groves of piccabeen and alexandria palms, pink toriga trees, strangler figs, vines, orchids and a prolific growth of ferns, including huge elkhorn ferns and many treeferns. In the cooler forests are mackay tulip oaks, silver quandongs and white beech, while the north-western section supports large, open eucalypt forests.

Walking Tracks

Many walking tracks begin at the Broken River campsite, ranging from short nature trails to long walks along the riverside to large rockpools, through rainforests to spectacular lookouts and Eungella township. Finch Hatton Gorge has tracks passing through a different type of rainforest to creeks and cascades. Experienced bushwalkers can explore the extensive wilderness areas of the park.

Climate

The park has a subtropical to tropical climate with a wet season from December to March or April, when mist and fog are common. The average rainfall is around 2000 mm. Summers are very warm and winters cool with frosts. Visit May to November.

Bookings and Fees

Campsites can be booked in advance, and permits must be obtained from the ranger. Fees are payable for camping and the use of facilities including guided tours and other activities.

Information and Bookings

The Ranger
Eungella National Park
Via Dalrymple Heights
Queensland 4741
Tel. (079) 58 4552

The Sphinx Rocks, one of the many striking granite boulder formations in Girraween National Park.

Left: *Overlooking the valley floor of Girraween National Park from a weathered granite peak.*

Girraween
National Park

Among the orchards and vineyards of the Granite Belt, high on the Great Dividing Range and just inside the Queensland border, the great granite outcrops of Girraween National Park rise up in a mass of tors, peaks and precariously perched boulders. This strange granite habitat is unique in Queensland and is the best wildflower area in eastern Australia.

Facilities
There are caravan and car camping areas at Bald Rock Creek and Castle Rock, both with hot showers, toilets, fireplaces and tables. Bush camping is permitted, but must be at least 3 km from roads and facilities. The visitor centre is at Bald Rock Creek, where there is also a public telephone and a picnic ground.

Location and Access
The park is about 260 km south-west of Brisbane, 32 km south of Stanthorpe and 15 km north of Wallangarra. Access is via the New England Highway, turn off on the sealed road to the information centre, camping and picnic grounds. Other roads within the park are unsealed.

Landscape
The park occupies 11 300 ha of rugged granite peaks rising to 1267 m at Mt Norman. Creeks flowing over granite beds have cut gullies and valleys through the park, and millions of years of erosion have sculpted the granite into dramatic shapes. Open eucalypt forests are established on the rock-strewn slopes, ridges and flats.

Flora and Fauna
Grey kangaroos, wallaroos, red-necked wallabies, long-eared brushtail possums, feathertail gliders, sugar gliders, greater gliders and wombats are among the mammals seen in the park.

Red-bellied black snakes, brown snakes, copper-tailed skinks, geckos, Lesueur's frog, the banjo frog and painted frog are occasionally encountered.

The forests and heaths support a diverse bird population including warblers, parrots, treecreepers, flycatchers and honeyeaters among the eucalypts, with robins, thornbills, wrens and firetails along the watercourses. Kookaburras, magpies, rosellas, wattlebirds and currawongs frequent the picnic grounds.

Twenty-five species of eucalypt have been identified in Girraween, they include the Wallangarra whitegum, New England blackbutt, round-leaved gum, yellow box and stringybarks. Girraween means 'place of flowers', and there are beautiful wildflower displays from July to October when

the heaths and shrublands burst into bloom. Wattles, pea flowers, daisy bushes and rock roses are common around the granite outcrops, with sundews and terrestrial orchids in swampy areas. Rainforest plants are found in moist gullies, including muttonwood, sweet pittosporum and possum wood.

Walking Tracks

Walking tracks lead to the tors and peaks, others to interesting natural formations and areas of great natural beauty. They vary from short nature walks to long climbs to the peaks. Drinking water should be carried on the longer walks.

Other Activities

There are many swimming spots in the river pools and some rock climbs for free climbers only on the granite outcrops; the ranger should be informed beforehand.

Climate

Winters are usually dry and cold with frosty nights, summers are warm with cool nights. Most of the rainfall of 850 mm occurs from November to March. Visit in spring for wild-flower displays, otherwise all year round.

The timid feathertailed glider is sometimes glimpsed in Girraween National Park.

Bookings and Fees

Bookings are essential during school holiday periods and long weekends for the campsites, and permits must be obtained from the ranger. No fees are charged.

Information and Bookings

The Ranger
Girraween National Park
Wyberba
Via Ballandean
Queensland 4382
Tel. (076) 84 5157

Hinchinbrook Island

National Park

Midway between Townsville and Cairns the massive mountain range of Hinchinbrook Island rises from a network of narrow channels separating Queensland's largest island from the coast. Permanent streams plunge down precipitous slopes into thick jungle, while long sandy beaches extend along the ocean side, patrolled by solitary sea eagles and washed by the warm waters of the Coral Sea.

Facilities

The two camping areas at Macushla and The Haven are reached by boat from the mainland. They have fireplaces, toilets and picnic tables, although Macushla has no fresh water, and visitors must bring sufficient to last their stay.

Bush camping is permitted although campers must move nightly. Accommodation is available at the island resort at Cape Richards.

Location and Access

The park is about 160 km north of Townsville via the Bruce Highway and the Townsville–Cairns rail link. Charter boats to the island can be arranged in Cardwell, it is 9 km to The Haven and 22 km to Macushla. Private boats can be launched from Cardwell, Lucinda and Dungerness. The resort has its own boat service from Cardwell.

Landscape

The park occupies 39 350 ha, comprising the whole of Hinchinbrook Island, a continental island of volcanic peaks rising to 1119 m at Mt Bowen, with dense tropical rainforests, eucalypt forests and impenetrable mangroves. The north-east part of the island is a low-lying strip of land with long beaches facing the ocean and dense mangroves towards the mainland, where brown mudflats appear at low tide. The jagged mountains forming the majority of the island remain one of the last untouched wilderness areas of the eastern seaboard.

Flora and Fauna

Bush rats, goannas and green tree snakes are common. In the forest the rich bird life includes yellow-breasted sunbirds, forest-kingfishers, green-winged pigeons, Torres Strait pigeons, sulphur-crested cockatoos, noisy pittas and scrub hens. Common birds of the mangroves include gerygones, honeyeaters, flycatchers, golden whistlers, and white-browed crake and striated heron.

The granitic soils of the island support mainly eucalypts with Moreton Bay ash, red bloodwood, forest red gum and white stringybarks. The tropical rainforests of the southern and western areas consist of such species as maples, white cedars, guandongs and fig trees, with many vines, epiphytic ferns and mosses. The mangroves are an important breeding ground and food source for crustaceans and other marine life.

Walking Tracks

Four walking tracks have been established; from the resort to Shepherd Bay through dense rainforests and eucalypts to the beach; from the Macushla campsite to Shepherd Bay; and a fascinating mangrove walk in Missionary Bay, where a raised boardwalk runs through thick mangroves to the sand dunes behind Ramsay Bay. Experienced bushwalkers can explore the natural wilderness areas.

Other Activities

Swimming, boating and fishing are very popular. There are boat trips from the resort to the reefs to view the coral.

Climate

Summers are hot and wet with occasional cyclones and swarms of dangerous sea wasps in the waters. Winters are warm. Visit from May to November.

Bookings and Fees

Camping permits must be obtained from the information centre and are available for one week at The Haven and Macushla. Fees are payable for camping and the use of facilities including guided tours and other activities. Bookings can be made up to 12 weeks in advance and not less than 6 weeks prior to visits.

Information

Officer-in-charge
Cardwell Information Centre
P.O. Box 74
Cardwell
Queensland 4816
Tel. (070) 66 8601

Overleaf left: *Forests extend to the foreshores along the sandy north-eastern coastline of Hinchinbrook Island.*

Overleaf right: *Sunset over the rock pinnacles in Lakefield National Park.*

Lakefield
National Park

Lakefield is Queensland's second largest National Park and the most accessible park on the Cape York Peninsula. An outstanding wildlife habitat, vast wetlands and lily lagoons, with a huge population of waterbirds, are found in this semi-wilderness area.

Facilities
Bush camping is permitted at a number of sites adjacent to rivers and waterholes in the park. A permit is required and can be obtained from rangers at Lakefield or New Laura. The ranger can also suggest a campsite which best suits your interests and vehicle capabilities. Visitors must be fully equipped for outback survival.

Landscape
A vast river system comprising main streams and tributaries of the Bizant, Normanby, Hann, Morehead and North Kennedy Rivers drains this 528 000 ha park. During the wet season these watercourses join to flood vast areas, eventually emptying into Princess Charlotte Bay. With the dry season, stream flow decreases leaving large permanent waterholes and lagoons, interspersed by dry sandy river beds.

Flora and Fauna
Most of the park is covered by grasslands and woodlands dominated by bloodwood, Darwin stringybark, Molloy red box with Moreton Bay ash, eucalypts and paperbarks spread out on the floodplains. Rainforest grows on the rich soil along the river banks while monsoonal scrubs are found on sections of the sandstone escarpments. Mangroves line mudflats along the coast and estuary shores. The swamps, creeks and lily-covered lagoons are home to thousands of brolgas, magpie geese, Burdekin ducks, cormorants, egrets and jabirus. Both the harmless freshwater crocodile and the potentially dangerous estuarine crocodile abound in the rivers and lagoons.

Location and Access
Lakefield National Park is located 320 km north of Cairns; 146 km north-west of Cooktown. The main access is from Cairns via the Peninsula Developmental Road to Laura then north to the New Laura ranger station and on to Lakefield ranger station. This is the only access for two-wheel-drive vehicles, however even this road can become impassable after rain. Check road conditions before leaving Cairns. Four-wheel-drive access from Cooktown is via Battlecamp and from Coen, via Musgrave and Saltwater Creek.

Walking Tracks
Because of its vast size there are no formal walking tracks in the park apart from a designated track at Kalpowar crossing.

Other Activities
Observing wildlife, boating down the rivers, photography and fishing can all be enjoyed within the park.

Climate
The best time to visit is during the dry season months of April to November. During April and May some late rain is possible and travel within the park may be restricted by boggy roads.

Bookings and Fees
Fees are payable for camping and the use of facilities including guided tours and other activities.

Information and Bookings
Queensland National Parks and Wildlife Service
Regional Centre
P.O. Box 2066
Cairns
Queensland 4870
Tel. (070) 51 9811

Ranger-in-charge
Lakefield National Park
PMB 29
Cairns Mail Centre
Cairns
Queensland 4871
Tel. (070) 60 3271

New Laura Ranger Station
Tel. (070) 60 3260

Lamington
National Park

The last large area of undisturbed subtropical rainforest in south-eastern Queensland is protected by Lamington National Park. A vast area of rugged mountains, waterfalls and panoramic views across sheer cliff walls and distant mountain peaks to the Pacific Ocean and Gold Coast, Lamington, with its cool, green rainforest, magnificent orchids and splendid bird life offers a striking contrast to the hot, humid coastal areas with their high rise buildings and busy streets.

Facilities
A car camping area is located in the Green Mountains area, with fireplaces, toilets, cold showers and a picnic area. A private camping area is located at Binna Burra. Bush camping

The eastern swamphen, a good swimmer, roosts at night in tree branches in south-eastern Queensland.

Moss and other moist rainforest vegetation thrive in the cool damp atmosphere of Lamington National Park.

is permitted, but must be more than 500 m from a graded track, 3 km from a camping area and 30 m from a watercourse. There are ranger stations at the Green Mountains area and Binna Burra.

Guesthouse accommodation is available at O'Reilly's at Green Mountains and Binna Burra Lodge, where basic provisions can be bought.

Location and Access
The Green Mountains area is 115 km south of Brisbane via the partially sealed road from Canungra. The Binna Burra area is 96 km south of Brisbane via Nerang and the sealed road from Beechmont. Access roads are not suitable for caravans.

Landscape
The 20 200 ha park covers a series of densely wooded ranges and valleys rising to more than 1100 m on the crest of the McPherson Range on the Queensland–New South Wales border. Many creeks and rivers originate in the park, and more than 500 waterfalls plunge into gorges and valleys through dense rainforest vegetation.

Flora and Fauna
Red-necked pademelons graze around the picnic grounds, while the forests are home to potoroos, bandicoots, pigmy possums, brush-tailed possums, gliding possums, marsupial mice and tiger-cats.

Bird life is prolific; there are crimson rosellas, satin bowerbirds, regent bowerbirds, brush turkeys, wrens, warblers, fantails, whipbirds, lyrebirds, king parrots, rainbow lorikeets, the elusive rufous scrub-bird and buff-breasted pitta.

In the rainforests are giant trees, the rough-barked hoop pine, giant stinging tree, cedars, beefwood and carribean trees. Flame trees and firewheel trees bring brilliant flashes of colour to the slopes. Orchids are common, and include the orange blossom, spider, olive, ravine and clumps of golden king orchids. On the high plateau are stands of ancient Antarctic beech trees whose limbs support aerial orchids.

Walking Tracks
Extensive walking tracks cross the park, passing forests, creeks and waterfalls to spectacular lookouts. There are self-guiding short nature trails, a senses trail for the blind, and full-day walks. The wilderness areas are perfect for experienced bushwalkers.

Other Activities
There are some challenging rock climbs, rockpools for swimming and great opportunities for nature study.

Climate
Summers are warm to cool and winters generally fine but often cold. Most rainfall is between November and March, and mists often surround the ranges. Nights are often cold.

Bookings and Fees
Camping permits and advice on sites must be obtained in advance. Fees are payable for camping and the use of facilities including guided tours and other activities.

Information and Bookings
The Ranger
Lamington National Park
Green Mountains Via Canungra
Queensland 4275
Tel. (075) 45 1734

The Ranger
Lamington National Park
Binna Burra
Via Nerang
Queensland 4211
Tel. (075) 33 3584

Lawn Hill National Park

Emanating from springs in the honeycombed limestone plateau of the Barkly Tableland in the north-western edge of Queensland, Lawn Hill Creek snakes through the ancient ramparts of the Constance Range on its way to the Gulf of Carpentaria. Sheer red, orange and grey cliffs rise up from the lush tropical vegetation surrounding the creek, a remote and enchanting oasis in the dry plains of northern Queensland, and the major attraction of Lawn Hill National Park.

Previous page: *Dense rainforests overhang the many waterfalls and gorges in Lamington National Park.*

Facilities

A camping area with fireplaces, showers and toilets has been established near the ranger station in the gorge. Picnic areas and a boat ramp are nearby. Ice, minor supplies and domestic pet accommodation are available at Adel's Grove, 6 km from the park entrance. The nearest petrol and accommodation are at Gregory Downs Hotel, 100 km east. Visitors must bring their own cooking fuel as firewood is not available in the park.

Location and Access

The park is 436 km north-west of Mt Isa via the sealed Barkly Highway, turn off at Camooweal on the unsealed road towards Burketown and follow the signs to the park, 248 km away. From Burketown the park is 200 km south-west on unsealed roads. During the wet season from December to March and after heavy rains the roads are impassable. Visitors should check road conditions before leaving, and carry sufficient fuel, water, food and vehicle spares. A dry-weather airstrip is located and serviced by Adel's Grove.

Landscape

The park covers 12 200 ha of mainly sparsely vegetated terraced plateau on the edge of the Barkly Tableland. Lawn Hill Gorge winds for some 3 km through the park, its walls rising 60 m above the creek. The gorge ends at the imposing Island Stack, a massive rock plug formed by the creek as it changes course. Two km upstream from the gorge end is a natural dam of crystalline limestone, a natural spa bath and waterfall.

Flora and Fauna

Agile wallabies are plentiful at the woodlands of the gorge, while solitary wallaroos, rock ringtail possums and bats live in the rocky areas. Freshwater crocodiles are common and should not be disturbed. Other reptiles include tortoises, water monitors and non-venomous olivine pythons.

Fairy martins build their mud nests beneath rocky ledges, shrike-thrushes and Australian kestrels are often seen.

The lush tropical vegetation around the creek includes cabbage palms, Leichhardt pines, paperbarks, figs and white cedar trees. On the plateau are scattered eucalypts and spinifex grasslands.

Walking Tracks

Some 17 km of walking tracks lead along the gorge around Island Stack to waterfalls, the spa bath and up to lookouts above the gorge. Bushwalkers can explore the area, but should be well prepared and inform the ranger of their plans.

Early morning reflections of the bare gorge walls in Lawn Hill Creek.

Overlooking Lawn Hill Creek, an oasis surrounded by rocky hummock grass plains.

Other Activities
The best way to explore the gorge is in canoes or inflatables. Motorised boats and fishing are prohibited. There are Aboriginal artefacts in the gorge including rock paintings and middens. Special boardwalk and viewing platforms have been constructed to preserve these sites for future generations.

Climate
The wet season extends from December to March, with storms during October and November, when daytime temperatures can be very high. Winter nights can be cold. Visit from April to November.

Bookings and fees
Camping permits must be obtained from the ranger, bookings must be made up to 12 weeks in advance. Fees are payable for camping and the use of facilities including guided tours and other activities.

Information and Bookings
The Ranger
Lawn Hill National Park
R.M.B. 12, M.S. 1463
Mt Isa
Queensland 4825

Lizard Island
National Park

In 1770 Captain Cook climbed to the highest peak on Lizard Island, now known as Cooks Look, to search for a passage through the outer ramparts of the Great Barrier Reef, through which he was able to steer the Endeavour *into deeper waters. Lizard Island is now a National Park fringed by a coral reef and featuring a superb sheltered lagoon with magnificent growths of coral and vivid green and blue giant clams.*

Facilities
Camping is permitted on the island where there are picnic tables, barbecue and toilet. There is a bush shower near the water well which is situated 300 m from the camping area. Aircraft forbid the carrying of stove fuel so the only fuel available is driftwood. A good first aid kit including antiseptic must be carried as coral cuts can be easily infected. A privately managed resort area offers accommodation and includes the use of glass-bottomed boats, snorkelling gear and boats and equipment for fishing trips, etc.

Location and Access
Lizard Island is located 95 km north-east of Cooktown off the north coast of Queensland. It may be reached by chartered boat or private vessel. Flights from Cairns via Cooktown operate every day except Thursday.

Landscape
The 1012 ha of Lizard Island National Park are virtually surrounded by a well-developed reef which loops around the sheltered 10 m deep lagoon. Grassland, patches of rainforest, shrubland, woodland and swamp comprise most of the land area with intermittent seasonal streams. Cooks Look (353 m) is the island's highest point.

Flora and Fauna
Sand beaches with casuarina and other fringing vegetation, shrubland, woodland, pandanus groves and mangrove swamp comprise the variety of habitats on Lizard Island. The island was named by Cook's party after numerous large dark sand goannas which still survive there. Ten species of lizards as well as five species of snakes, all harmless, exist on the island. Abundant bird life number some 40 species, 14 of which are sea birds.

Walking Tracks
A marked track leads from Watsons Bay to the top of Cooks Look. Elsewhere on the island easy bushwalking enables the visitor to observe the park's scenery and wildlife.

Other Activities
The surrounding reef offers excellent opportunities for swimming, diving and snorkelling. A range of recreational activities are provided for resort guests.

Climate
Cyclones may occur from November to April and it is the wet season from December to March. From May to October there are strong south-easterly winds.

Bookings and Fees
Fees are payable for camping and the use of facilities including guided tours and other activities. Camping permits are required. The camping site will accommodate only about 12 people. Bookings can be made up to 12 weeks in advance through the Cairns office of the Queensland National Parks and Wildlife Service. Bookings for the resort can be made through the Queensland Government Travel Centre.

Information and Bookings
National Parks and Wildlife Service
Queensland Regional Centre
P.O. Box 2066
Cairns
Queensland 4870
Tel. (070) 51 9811

Overleaf: *Coral reefs protect the blue lagoon on Lizard Island.*

Mount Spec National Park

The summits and escarpments of the Paluma Range rising almost 1000 m above the floodplain of Big Crystal Creek are patched in various hues of green reflecting the many different types of forest which have devleoped in Mt Spec National Park. Beneath are waterfalls, cascades and swimming holes shaded by the magnificent rainforested valleys which offer excellent leisure opportunities and spectacular scenic viewing.

Facilities

The campsite is located on Big Crystal Creek near the Paradise waterhole. Facilities include showers, toilets, barbecues and tables. All campers require a permit from the Queensland National Parks and Wildlife Ingham information office. Several picnic areas nearby have barbecues and tables.

At Little Crystal Creek and McClellands Lookout toilets, tap water, tables and barbecues are also provided. Bush camping is permitted at the ranger's discretion and permits are required by all campers.

Location and Access

The park is situated 70 km north-west of Townsville turning west off the Bruce Highway on to a narrow bitumen road passing Little Crystal Creek, Paluma and McClellands Lookout. A dirt road to Big Crystal Creek leaves the highway 2 km north of the Paluma turnoff.

Landscape

Mount Spec National Park is an area of 7724 ha straddling the Paluma Range. The deeply dissected escarpment reaches down towards numerous creeks with many waterfalls and rock-pools. Heavy rainfall on the summits averaging 2850 mm per annum gives rise to dense rainforest areas often shrouded in mist, while the lower areas of the park comprise drier eucalypt forests and riverine vegetation.

Flora and Fauna

Along the park's western boundary a belt of rainforest extending through from Townsville to Cooktown forms an imposing buttressed canopy including such species as Queensland maple, teak and hoop pines. An understorey of vines and ferns and a deep layer of leaf litter provide habitats for a number of mammals, most commonly green ringtail possums, white-tailed rats and pademelons.

Further down the range, ironbarks, bloodwoods, poplars and the kapok tree are found while along the creeks vegetation is comprised of casuarinas and banksias, a home for doves, kookaburras and agile wallabies. Bird life is prolific and includes brush turkeys, Lewin's honeyeaters and in the northern rainforest, the golden bowerbird, the tooth-billed catbird and the Macleay's honeyeater.

An old bridge passes over cascades on Crystal Creek in Mount Spec National Park.

Walking Tracks

At McClellands Lookout a track of about 2 km links the picnic grounds with a series of lookouts. Marked tracts lead to Witt's Lookout and Cloudy Creek. Longer tracks that were once the pack trails of early twentieth-century tin miners lead north and west through the range. The ranger should be notified of any intended bushwalks.

Other Activities

Swimming is very popular along Crystal Creek while McClellands Lookout, to the south of the park, affords panoramic views of Crystal Creek and Halifax Bay.

Climate

Summers are warm and wet and winters are cold and usually dry. Autumn and spring are the most pleasant times to visit Mount Spec.

Bookings and Fees

Fees are payable for camping and the use of facilities including guided tours and other activities. A permit for camping may be obtained from the Ingham office of the Queensland National Parks and Wildlife Service

Information

Ranger
Mount Spec National Park
C/o Main Street
Paluma
Queensland 4816
Tel. (077) 70 8526

Queensland National Parks and Wildlife Service
2 Herbert Street
P.O. Box 1293
Ingham
Queensland 4850

Victoria

Victoria is Australia's smallest mainland state and the most densely populated and intensively farmed. Despite this, nearly one-third of natural bushland remains. The National Parks reflect the wide range of Victoria's landforms; alps, open grasslands, desert mallee, tropical fern gullies, ash forests and volcanic plains. Most of the country is mountainous; the Great Dividing Range looms across the state, its outlying ridges pushing to the Bass Strait coast.

National Parks in the north-east preserve the lofty plains of the Victorian Alps, snow-covered in winter, but dotted with wildflowers at other times. Here, Mt Buffalo National Park draws skiers for downhill and cross-country skiing. Another winter haven is Baw Baw National Park, ideal for ski-touring and mountain climbing.

Croajingolong, on the north-east coast of the state, is Victoria's largest park and comprises rainforest, woodland, heath, rocky promontories and ocean beaches. Other parks in the east include temperate rainforests and subtropical remnants at Morwell, Tarra Valley and Bulga, and an outstanding variety of plant communities around the granite peaks of Wilsons Promontory.

From the Port Campbell coast, northwards to the Murray River, the landscape is one of great contrasts. Ranges in the west are backdrops for remarkable coastal scenery at Otway and Port Campbell National Parks. Perpendicular cliffs, deeply eroded by wind and water, and masses of spring wildflowers are found in the Grampians National Park; while the Little Desert National Park occupies remote sandplains and mallee country. Wyperfeld National Park, in the arid north-west corner of the state has unique mallee flora and wildlife including the mallee fowl. In the far north, the Hattah-Kulkyne National Park is an oasis of leafy woodlands, wetlands and lakes among the semi-arid grasslands and sand ridges of the area.

Spacious areas are reserved for National Parks in the highlands fringing Melbourne itself. Parks here, such as Ferntree Gully and Kinglake, are particularly attractive to bushwalkers. Victoria's richest display of spring wildflowers are seen at the Brisbane Ranges National Park, little more than an hour's drive from Melbourne. In summer, many bushwalkers prefer the parks along the coast, such as the Nepean parks, Cape Schanck, The Lakes and Wilsons Promontory.

Spring and summer are the best times to visit Victoria's National Parks, when wildflowers bloom in profusion and sun lovers can head for the parks along the coast to swim, surf, canoe, boat or fish. Autumn, with its mild weather, beckons the bushwalkers and winter means skiing at the beautiful alpine parks.

Baw Baw National Park

The granite Baw Baw Pateau was crossed by Victoria's first long-distance walking trail, established between Warburton and Walhalla in 1906; a portion of this track is now part of the Alpine Walking Track. Unfortunately the disastrous fires of Black Friday on 13 January 1939 burnt most the area traversed by the walking track. The plateau offers outstanding views, colourful wildflowers, extensive snowplains and heathlands and, in complete contrast, steep river valleys with dry open forests. In winter the park is popular for ski-touring and in summer bushwalkers enjoy the fine scenery.

Facilities

There are pit toilets, picnic facilities and fireplaces at the Mt Erica car park, Eastern Tyers River and at the Aberfeldy River campsites where bush camping is also permitted. The St Gwinear car park has picnic facilities, fireplaces and toilets as well as food and ski hire service during the winter months only. Vehicle-based camping is not permitted in the section of the park which is within the Thomson Dam catchment.

The park office is situated at the Department of Conservation, Forests and Land Depot at Erica.

Location and Access

The park is located 180 km east of Melbourne, north of the La Trobe valley. Access is off the Princes Highway via Thomson Valley Road north of Erica, the last 13 km of the road to St Gwinear is unsealed.

Within the park the St Gwinear road is hazardous during winter and chains should be carried. The road past Walhalla and down into the Aberfeldy area is unsuitable for caravans and many of the tracks in the Thomson–Aberfeldy area are suitable for four-wheel-drive vehicles only.

Landscape

The Baw Baw National Park takes in most of the Baw Baw Plateau, a mountainous block at the southern end of Victoria's high country, as well as sections of the Thomson and Aberfeldy river valleys. Mt St Phillack, the highest peak, reaches 1566 m. Above 1300 m snow covers the ground from July to September. Extensive treeless tracts of subalpine bog and peaty swamps form on poorly drained, lower-lying ground where a number of major rivers begin their courses to steep, heavily forested gullies at the edge of the 13 000 ha park.

Flora and Fauna

Altitude, aspect and soil types have created great botanical diversity within the park. It contains a complex of subalpine vegetation and a range of open forest types in which alpine ash, mountain ash, messmate, red stringybark, and silvertop are common. Notable species in the subalpine woodlands include Tingaringy gum, myrtle beech, mountain plum pine and Baw Baw berry, one of two ericas native to Victoria. An extremely rare clubmoss is found on the plateau.

Rich mosses grow profusely in the moist gullies of Baw Baw National Park.

Fauna within the park is of global significance. Leadbeater's possum and the Baw Baw frog are recognised endangered species. Other more common animal life includes wombats, wallabies, echidnas, possums, platypuses and gliders. Reptiles in the park include tiger, black, copperhead and white-lipped snakes and there are also several lizard species, including goannas. Among the common birds are superb lyrebirds, yellow-tailed black cockatoos, gang-gangs, crimson rosellas, flame robins and ravens.

Walking Tracks

Short walking trails into the park meet a 20 km section of the Alpine Walking Track, crossing the plateau from north to south past Mt St Phillack and Mt Erica. Short walks include the 1 km Beech Gully nature walk and tracks on to the plateau from Baw Baw Alpine Reserve and the St Gwinear and Mt Erica car parks. Longer walks involve overnight camping on the plateau.

Other Activities

Ski-touring is one of the major attractions of Baw Baw National Park. During winter many summer hiking tracks become cross-country ski-touring routes. A pole line leads from St Gwinear car park to Baw Baw Village and temporary ski-pole lines are erected during winter. A winter information brochure is available from the park. Baw Baw Village and the downhill skiing areas adjoin the park.

Outstanding sightseeing, canoeing and trout fishing (a licence is required) are also enjoyed. Horse riding is restricted to four-wheel-drive tracks only, and is not permitted on the plateau.

Climate

Visit year round. Severe weather changes can occur suddenly and ski-tourers and bush campers must be adequately equipped to cope with emergencies.

Bookings and Fees

There are no fees payable and bookings are not required, however it is always wise to enquire during holiday periods.

Information and Bookings
Ranger-in-charge
Baw Baw National Park
School Road
Erica
Victoria 3825
Tel. (051) 65 3204

Croajingolong
National Park

Set in some of Victoria's most beautiful lakeland, and extending along the coast some 100 km south from the New South Wales border, Croajingolong ranks among the finest National Parks in Australia. The park preserves the bays, islands and lakes of Mallacoota Inlet; long, deserted beaches, vast dune systems, rocky promontories and flooded river valleys backed by forested ridges and steep-sided gullies.

Facilities

The two main car camping areas are located at Wingan Inlet and Thurra River, with fireplaces, toilets, water and picnic tables. Small boats can be launched at Wingan Inlet, but the road is not suitable for caravans. At Shipwreck Creek and Mueller River there are car campsites with toilets. Bush camping is permitted, but the ranger should be consulted about suitable sites. Park offices are at Cann River and Mallacoota.

A range of accommodation is available in Mallacoota.

At the Thurra River estuary water reeds are met by high coastal dunes.

Right: *Magnificent sunrise over the popular boating and fishing resort at Mallacoota Inlet.*

Location and Access

Mallacoota Inlet is about 560 km east of Melbourne via the Princes Highway. Unsealed access roads from the Princes Highway lead to Wingan Inlet (West Wingan Road), Thurra River and Mueller River (Tamboon Road). Shipwreck Creek is accessed via the unsealed Betka Road from Mallacoota.

Landscape

The park covers 86 000 ha extending in a coastal strip some 10 km wide. Many creeks flow through the park to the ocean, originating in the South Coast Range and winding through the low, coastal plains which rise to 490 m at Genoa Peak north-west of Mallacoota. A low mountain range dominates the skyline to the north-east of Mallacoota Inlet, while sand dunes cover most of the coast. The drowned river valleys forming Mallacoota and Wingan Inlets are fringed with forests, while sand dunes cover pink granite rocks along the coast which jut out in prominent headlands in places.

Flora and Fauna

Grey kangaroos, brushtail and ringtail possums, feathertail and sugar gliders are common. Australian fur seals and little penguins inhabit the islands off the entrance to Wingan Inlet. Bird life is prolific and includes many black swans, pelicans, crested terns, cormorants, grebes and dotterels on the lakes, with lyrebirds, kookaburras, wattle birds and eastern whipbirds in the forests.

The park is renowned for its range of flora. There are rainforest gullies with such dominant trees as the lillypilly, sweet pittosporum, mutton-wood and leatherwood. Ferns and orchids are abundant. There are open eucalypt forests with brown and yellow stringybarks, yertchuk and mountain grey gums. Around the inlets are red bloodwood trees, silvertop ash, swamp paperbarks, banksias and wattles.

Dense vegetation cloaks the banks of the creeks in Croajingolong National Park.

Left: *Vast sand dune systems border the south-eastern coastline in Croajingolong National Park.*

Walking Tracks

A variety of bush, beach and river walking tracks have been established, ranging from short nature trails around the camping areas to long walks along rivers and lakesides, through heathlands, forests and sand dunes. There are many wilderness areas, particularly in the eastern section of the park, with attractions for experienced bushwalkers.

Other Activities

Swimming, fishing canoeing and surfing are popular activities. Boats can be hired at Mallacoota to explore the lakes.

Climate

The park has warm summers and cool winters. Visit all year round.

Bookings and Fees

Advance bookings are required for Christmas, Easter and school holidays; bookings are not accepted for other periods. Camping fees are charged.

Information and Bookings

Ranger-in-charge
Croajingolong National Park
Princes Highway
Cann River
Victoria 3889
Tel. (051) 58 6351

Fraser National Park

Fraser National Park, on the western shores of the man-made Lake Eildon, is a sublime retreat among the forested ridges and peaks of the foothills of the Victorian Alps. The numerous indented bays and long, narrow spurs plunging into the lake came into existence by the creation of Eildon Reservoir in the 1950s. Subsequent reclamation of land cleared by local graziers and gold diggers to form the park has encouraged regeneration of the eucalypt forests and the return of native wildlife.

Facilities

The three car and caravan camping areas around the shores of Coller Bay at Devil Cove, Lakeside and Candlebark have fireplaces, toilets, hot showers and laundries. The park headquarters, picnic areas and launching ramps are also located at Coller Bay. Bush camping is permitted in consultation with the ranger.

Accommodation and supplies are available at Alexandra and Eildon.

Location and Access

The park is 145 km north-east of Melbourne and 17 km from Alexandra. Turn off the Maroondah Highway at Alexandra towards Eildon and follow the road into the park. Roads within the park are sealed.

Landscape

The park occupies 3750 ha of hills and ridges along the indented shoreline of Lake Eildon. The western boundary follows the sandstone ridge of the lower Puzzle Range, with panoramic views over Coller Bay and the surrounding peaks. Steep spurs run from the ridge to the lake with creeks following the valleys between. Around the lakeside are large, open, grassy areas, while on the hillsides are forest patches.

Flora and Fauna

Eastern grey kangaroos and brush-tailed possums are abundant, swamp wallabies are sometimes seen, as are gliders, ring-tailed possums, echidnas and marsupial mice.

Around the campsites are crimson rosellas, black-backed magpies, noisy miners, white cockatoos, galahs and kookaburras. Other common species of the woodlands include gang-gang cockatoos, fantails, robins, rufous whistlers,

Right: *Water sports and lake-shore camping and picnicking are popular activities in Fraser National Park.*

Early morning tranquillity over Lake Eildon in Fraser National Park.

honeyeaters and thornbills. There are black and pied cormorants, pelicans, ducks, black swans, white ibis, white-faced herons and black-fronted dotterels around the lake.

Common trees include broad-leaved peppermints, red stringybarks and red box, with candlebark gums, and yellow and long-leaved box trees in restricted areas. Common shrubs include silver wattles, the cherry ballart, white-flowered cassinia and yellow-flowered guinea flower. Clustered everlastings and ground orchids are common.

Walking Tracks
There are a number of walking tracks, including a nature trail near Devil Cove and four short walks beginning at Keg Spur Gate near Coller Bay. Long walks ascend the ridges and spurs to beautiful lookouts, or follow the lakeside.

The brilliant red and blue plumage of the male gang gang cockatoo adds colour to the landscape in Fraser National Park.

Other Activities
Swimming, trout fishing, canoeing and boating are popular activities. There are a number of old goldmining shafts and relics in the park.

Climate
The park has warm summers and cold winters. Visit in spring, summer and autumn.

Bookings and fees
Campsites must be booked during peak holiday periods, and permits obtained from the ranger. Camping and park entry fees are charged.

Information and Bookings
Ranger-in-charge
Fraser National Park
P.O. Box 153
Alexandra
Victoria 3714
Tel. (057) 72 1293

Hattah-Kulkyne National Park

The periodic flooding of the Murray River has created an oasis of leafy woodlands amidst the sand ridges and grasslands of north-western Victoria. Here, on 49 550 ha south of Mildura, the Hattah-Kulkyne National Park provides an outstanding waterbird habitat and an ideal location for camping, bushwalking, canoeing and nature study.

Facilities
There are two designated camping sites, one at Lake Hattah and the other on the shores of Lake Mournpall. Picnic tables, fireplaces and toilets are provided as well as water at Lake Hattah, however supplies are limited and visitors are advised to carry water. There is an information centre at Lake Hattah providing a guide to activities and facilities.

Vehicle-based bush camping is not permitted and bushwalkers planning overnight trips must inform the ranger in charge.

Supplies are available at Hattah and Nangiloc.

Location and Access
Hattah-Kulkyne is located 500 km north-west of Melbourne and may be reached by the Murray Valley Highway via Wemen on a sealed road as far as Lake Hattah and an unsealed track to Lake Mournpall. Alternatively, motorists may take the Calder Highway from Mildura via Hattah.

Landscape
Hattah-Kulkyne National Park features lakes, billabongs and large areas of mallee woodland. Floodwaters from the Murray River spill west from Chalka Creek to replenish a system of lakes including the Hattah, Arawak, Bulla, Marrramook and Nip Nip.

Flora and Fauna
Plant species along the watercourses and river flats include black box, river red gum and native pine. By far the largest proportion of Hattah-Kulkyne is mallee scrub. Wildflowers are abundant in spring.

The area supports unique Australian fauna—red kangaroos, western grey kangaroos, possums, echidnas, emus and the mallee fowl. On the lakes, black swans, pelicans, grey teal, mountain duck and many other duck species may be

found while around the lake and swamp margins, white egrets, herons, spoonbills and ibis can be seen. The hollow branches of the big river red gums are nesting and roosting places for countless galahs, white and pink cockatoos and many species of parrots.

Pelicans find a luxurious breeding ground at Lake Hattah.

Walking Tracks

A self-guiding trail starting near the park entrance offers a leisurely two-hour walk or a 30-minute drive as an introduction to the park. There are also numerous other interesting walking trails through the countryside.

Other Activities

Hattah-Kulkyne is a superb location for birdwatching and photography. Swimming and boating are also popular.

Climate

The park is best visited in spring, autumn or winter as summer is extremely hot.

Bookings and Fees

No entrance or camping fees are payable and bookings are not required.

Information

Park Office
Hattah-Kulkyne National Park
R.S.D. Hattah
Victoria 3501
Tel. (050) 29 3253

The delicate pink and white plumage of the Major Mitchell cockatoo is a frequent sight at Hattah-Kulkyne National Park.

Kinglake
National Park

The largest National Park near Melbourne, Kinglake is an area of cool, fern-filled gullies, wooded valleys and timbered ridges which provide a perfect setting for two breathtakingly beautiful waterfalls, Masons and Wombelano Falls. The park is a favourite haunt for bushwalkers.

Facilities
There are camping and caravan sites with fireplaces, toilets and water available at the Gums in the northern section of the park. Mason Falls, Frank Thompson Reserve and Jehosaphat Gully picnic areas have tables, fireplaces, water and toilets. There is also a picnic area on Eucalyptus Road and limited barbecue facilities are available at the Steels Creek picnic ground. The park office is located on National Park Road.

Location and Access
Kinglake National Park is located 65 km north-east of Melbourne. Access is off the Yarra Glen–Yea Road and Whittlesea-Yea Road at Kinglake West.

Landscape
The park takes in 11 270 ha of eucalypt forest on the slopes of the Great Dividing Range where it curves closest to Melbourne. North-south spurs, 500–600 m high, are separated by steep, damp valleys of streams that drop sharply to the lowlands before joining the Yarra River. From a lookout point visitors can take in a panoramic view of the Yarra Valley, Port Phillip Bay and the You Yangs Range.

Flora and Fauna
The park is rich in plant life with tall open forest including mountain ash, mountain grey gum, messmate, narrow-leaf peppermint and blackwood. The understorey includes blanket-leaf, musk daisy bush, Victorian Chrismas bush and more than 40 identified species of ferns, testifying to the moist climate and altitude of the area. Red stringybark, broad-leaved peppermint and messmate stringybark dominate the ridge tops. Here, the groundcover is varied and colourful during the spring and early summer when plants such as heath, prickly moses, hop goodenia, bitter-pea and bush-pea burst into bloom.

The moist gully habitat is ideal for platypuses and water rats. Also seen are the swamp wallaby, eastern grey kangaroo, koalas and echidnas. Over 100 species of birds have been identified in the park. Lyrebirds are numerous, but not often seen; the males can be heard practising their mimicry in winter. Common species of small birds found in the wet gullies include grey shrike-thrushes, superb fairy wrens, scrubwrens, eastern yellow robins and the white-throated treecreepers which cling to the bark of peppermints and messmates searching for food. Larger birds such as currawongs, kookaburras, crimson rosellas and noisy yellow-tailed black cockatoos are regular visitors. The eastern spinebill and white-eared honeyeaters can be found amongst the banksias and heath.

Walking Tracks
There are many tracks in the park providing for walks of various length and difficulty. Some have steep pinches, but they command excellent views over Melbourne and the surrounding ranges. The Everard Circuit Walk enables the serious bushwalker to undertake the challenging 6-hour circuit walk close to Melbourne. The walk passes through the major habitats of the park and provides fine views.

Other Activities
Bushwalking is the major activity in the park but picnicking, birdwatching, scenic views and horse riding can all be enjoyed. During some weekends and holidays the rangers organise special family fun activities; contact the park office for details.

Climate
Spring and autumn are the best times to visit.

Bookings and Fees
Camping fees are payable at the Gums campsite and bookings are required for the Christmas–January, Labour Day, Easter and Melbourne Cup holiday periods.

Information and Bookings
The Ranger
Kinglake National Park
National Park Road
Pheasant Creek
Victoria 3757
Tel. (057) 86 5351

The moist gullies in Kinglake National Park are covered in lush ferns.

The fascinating mallee fowl build nesting mounds all over Little Desert National Park.

Little Desert
National Park

The Little Desert National Park is a desert in name only. To the early farmers unable to grow their crops in its poor soils, the Little Desert did appear desert-like, but to visitors today the area is full of life. Annual rainfalls of 400 mm support a variety of vegetation types; wildflowers are particularly abundant in the heaths during spring. The area also provides habitats for a variety of wildlife including the unique mound-building mallee fowl, and the park is of great interest to all natural history enthusiasts.

Facilities
Picnic tables, fireplaces and toilets are provided at the picnic area and campground (both south of Kiata). The water supply is limited and campers are asked to bring their own water. Bush camping is permitted by arrangement.

Location and Access
The park is located 365 km north-west of Melbourne; 10 km south of Kiata. The northern part of the park is best approached from Kiata along the Western Highway; a sealed road leads to picnic and camping areas and walking tracks. The eastern part of the park can be approached along a gravel road from Dimboola which provides access to the Wimmera River and walking tracks. A sealed road from Nhill leads to a nature walk through heath and woodland.

Landscape
Sandy plains extend west through the 35 300 ha park from woodlands on the alluvial Wimmera River floodplains. The major soils in the park are wind-blown sands that support mostly stringybark mallee and heath plants. Scattered throughout sandy areas of the park are iron-rich sandstones called laterite. A salt lake is situated near the centre of the park.

Fires, both naturally occurring and caused by man, have long played an important role in the Little Desert. Charcoal stumps and burnt bare patches of ground are evident in many parts of the park for months and often years after a fire has swept through. A large wildfire occurred in 1977 when most of the eastern part of the park was burnt.

There are remnants of a few Aboriginal transit camps along the river floodplains of the park, where axeheads, weapons and mill stones have been found.

Flora and Fauna
Mallee and heath plants grow on the wind-blown sands of Little Desert National Park. The heathlands are spectacular in spring with splashes of pink and red brush and flame heath contrasting with the rich yellow of wattle and guinea flowers. Subtle colour is further added to this display by the masses of white flowers of the beard heath and fringed heath-myrtle. Smooth-barked yellow gums grow on isolated clay pans, and patches of reddish soils among the sand dunes foster broom-brush and stringybarks. Along the eastern boundary of the park the alluvial floodplain soils support river red gum and black box woodlands.

There is great variety of bird life with over 200 native birds species having been sighted in the park. Parrots, wrens and currawongs are common, and the yellow gums provide a prolific source of nectar for babblers, wattlebirds and honeyeaters. Perhaps the most fascinating bird found here is the mallee fowl—an abandoned nesting mound of sand and litter can be seen on one of the nature trails. Brush-tailed possums, kangaroos and bats, as well as emus and many different species of reptiles are found here. Bearded dragons and stumpy-tailed lizards can often be seen basking in the sun.

Walking Tracks
Several nature trails have been developed to sample the park's distinctive character.

Other Activities
Birdwatching, photography and observing the wildflowers are popular activities within the park.

Climate
Visit during spring (for the wildflowers) and autumn. The summer heat can be quite harsh.

Bookings and Fees
Camping fees are charged; bookings are not required.

Information
Ranger-in-charge
Little Desert National Park
R.M.B. 389
Nhill
Victoria 3418
Tel. (053) 89 9218

Lower Glenelg
National Park

In the south-west corner of Victoria the Glenelg River flows through a deep gorge, etched through the limestone plateau of this ancient seabed for the final 24 km of the 400 km journey from the Grampians to the Southern Ocean. The Lower Glenelg National Park protects 20 km of the Glenelg Gorge, together with a number of limestone caves where the remains of such long extinct animals as the giant kangaroo and marsupial lion have been found. The park is a refuge for wildlife displaced by extensive land clearing in the district, and a centre for boating and fishing enthusiasts.

Facilities

Car and caravan camping areas are located at Pritchards and the Princess Margaret Rose Caves where there are fireplaces, toilets and picnic tables. The latter also has three motor cabins for rent by families and a kiosk. Many bush camping sites along the Glenelg River are accessible by car and have toilets and picnic tables. Several boat ramps and picnic areas have been provided along the river, and there is an information centre at the park headquarters in Nelson.

A range of accommodation is available at Nelson.

Location and Access

The park covers 27 300 ha around the Glenelg River Gorge, 40 to 50 m deep in places, flanked by sheer cliffs and surrounded by eucalypt forests. Kentbruck Heath in the eastern section of the park is noted for its wildflowers growing on sandy ridges and gravelly rises, drained by a number of small creeks. The Princess Margaret Rose Caves in the north-west of the park have spectacular and colourful crystalline formations and vertical shafts: death traps for unwary animals.

Flora and Fauna

Eastern grey kangaroos, red-necked wallabies, brush-tailed possums, potoroos, the swamp antechinus, heath rats and echidnas are numerous in the park. There is one remaining colony of wombats and a few yellow-bellied gliders. Bats congregate in the caves; platypuses and koalas may be seen.

Emus, herons and ducks are common. There are many gang-gang cockatoos, southern emu wrens, thornbills, honeyeaters, fantails, flycatchers, quails, the azure kingfisher and the rare rufous bristle bird.

The park has flora from both the east and west of Australia. Eucalypt forests cover large areas, with such species as messmates, brown stringybarks, manna gums and river red gums. In the gullies are soft treeferns, pea flowers, Mitchell wattles, daisy brushes, hop goodenias and wiry baueras. The heaths bloom with wattles, boronias, daisy flowers, pea flowers, epacrids, correas and ground orchids.

The River View Nature Walk reveals the meandering passage of the Glenelg River.

Left: *Lush native flora creates a picturesque border along the Glenelg River.*

Bizarre crystalline formations in Princess Margaret Rose Caves.

Walking Tracks
A number of walking tracks run through the park and from the road to the river. An interesting nature trail has been established near the caves, there are riverside walks and tracks through the forest and heath. The 250 km Great South West Walk passes through the park.

Other Activities
The Glenelg is an excellent river for canoeing, fishing and swimming. Water skiing and power boating are permittted in designated areas.

Climate
The park has warm summers and cool winters. Visit all year round.

Bookings and Fees
Campsites can be booked in advance, permits are required and fees charged.

Information
Ranger-in-charge
Lower Glenelg National Park
C/o Post Office
Nelson
Victoria 3292
Tel. (087) 38 4051

The shy echidna or spiny anteater is sometimes seen in Lower Glenelg National Park.

Morwell
National Park

Magnificent forest scenery, resounding with bird calls, soft fern gullies and rare orchids can be found in the small, but beautiful, Morwell National Park. Rapid subdivision and development in the 1960s prompted the former owner of Morwell to offer it for conservation—the area was not totally untouched, however, it contains one of the last patches of natural Gippsland forest.

Facilities
No camping is permitted; accommodation is available at Churchill or Morwell. Picnic tables, fireplaces, drinking water and toilets are provided at the Kerry Road picnic area, and there are also picnic facilities at Lyndons Clearing.

Location and Access
Morwell National Park is located 165 km south-east of Melbourne; 12 km south of Morwell. Access is off the Midland Freeway 2 km past Churchill.

Landscape
Situated in the northern foothills of the South Gippsland highlands, the 404 ha park comprises steeply sloping forested ridges and fern gullies. It is one of the last existing patches of Gippsland forest.

Flora and Fauna

The vegetation ranges from dry sclerophyll forest in the west to heavily timbered fern gullies with tall mountain grey gums and blackwood trees. Over 40 species of orchids are found, from the tiny bird orchids of autumn to the summer-flowering hyacinth orchid, one of the largest ground orchids in Victoria. In particular, the park is the habitat of the rare butterfly orchid, *Sarochilus australis,* a tree-clinging plant that in summer sends out a drooping stem clustered with yellow-green flowers distinguished by a tongue of red and yellow markings. The orchid grows on sheltered trees in the gullies.

The park is rich in bird life with over 100 identified species. Yellow robins, grey shrike-thrushes and crimson rosellas are often seen and superb lyrebirds can be heard. Koalas, wallabies, kangaroos, possums and many small marsupials are common. Reptiles seen in the park include the black, copperhead and tiger snake and several lizard species including goannas.

Walking Tracks

The park has several pleasant, easy walking tracks. Fosters Gully nature walk provides an introduction to the natural features of the park and is an enjoyable walk for families. Stringybark Ridge track leads through the drier sections of the park. Fine views can be obtained from Jumbuck Road which forms part of the park's eastern boundary.

Other Activities

General sightseeing, bushland relaxation, birdwatching, photography and observation of flora can all be enjoyed within the park.

Kangaroos are only one of the many animals that abound in Morwell National Park.

Climate

Spring, summer and autumn are the best times to visit. The wildflowers are best seen in spring and early summer.

Bookings and Fees

There is no camping in the park.

Information

Ranger-in-charge
Morwell National Park
P.O. Box 19
Churchill
Victoria 3842
Tel. (051) 22 1478

Mount Buffalo National Park

Towering above the Ovens Valley, the high granite plateau of Mount Buffalo National Park looks out over the spectacular peaks of the Victorian Alps. The sheet cliffs and eroded monolithic outcrops resembling a great humped buffalo resisted attempts at exploration for many years. Yet today this is one of Victoria's most popular National Parks, with endless walking tacks, crystal clear mountain streams, delicate wildflowers and excellent skiing and tobogganing in the winter snows.

Facilities

The camping area at Lake Catani has access for cars, but not caravans, with fireplaces, toilets, hot showers and a laundry. The park office, visitor centre and chalet are nearby. The chalet has basic supplies, meals , a telephone, post office, ski hire and guest house accommodation (booked through the Victorian Railways). Tatra Inn in the south of the park has basic supplies, meals, petrol and accommodation (booked through the Victorian Government Tourist Bureau). Bush camping is allowed, but must be well away from roads and facilities. Snow camping is permitted, but no more than 20 campers are allowed in the camping area for a maximum of two nights.

Location and Access

The park is 335 km north-east of Melbourne via the Hume Highway and Ovens Highway, turn off at Porepunkah on the steep, sealed road to the park. The main access roads within the park are sealed, snow chains must be carried in winter, and radiators require anti-freeze.

Landscape

The park covers 31 000 ha, encompassing the Mount Buffalo Plateau, rising to 1724 m at The Horn in the south, and 1539 m at Mt McLeod in the north. The plateau consists of granite tors, rounded boulders and great rock formations like The

Monolith, Torpedo, Cathedral and Egg Rock. Numerous creeks originate in the park, plunging down in spectacular waterfalls over sheer cliffs to the valleys below. There are two lakes, Catani and the Reservoir.

Flora and Fauna

Swamp wallabies, wombats, possums, yellow-bellied gliders, antechinus and bats inhabit the park. Lyrebirds are often seen, as are gang-gang cockatoos, honeyeaters, robins, pardalotes, fantails, thornbills and crimson rosellas.

On the lower slopes are eucalypt forests with stringybarks and peppermints. Higher up and on the plateau are woodlands dominated by snow gums, alpine ash, mountain gums and the unique Buffalo sallee. There are large heathlands with low shrubs, sedges and heaths, and beautiful summer wildflower displays with such species as alpine oxylobium, leafy bossiaea, royal bluebells, royal grevillea and everlastings.

Walking Tracks

More than 140 km of walking tracks have been constructed, from nature walks to long walks to spectacular viewpoints, waterfalls, peaks and woodlands. Rangers conduct nature walks and animal spotlighting excursions in holiday periods.

Left: *Spring and summer visitors to Mt Buffalo National Park may climb the rocky pyramid known as The Horn (1700 m).*

Mist enshrouds the deep gorges at Mount Buffalo, Victoria.

Buffalo House, a grand old ski lodge at Mt Buffalo built in the 1920s.

Other Activities

Swimming, sailing and canoeing are popular on Lake Catani. There are ski lifts at Dingo Dell and Crests, a number of toboggan runs and cross-country skiing trails.

Climate

The park has warm summers with cool nights and cold winters. Warm clothing is needed all year round. Visit in summer for wildflower displays and winter for skiing.

Bookings and Fees

Bookings are required for campsites in holiday periods; permits must be obtained from the ranger and fees paid. A park entry fee is charged.

Information

Ranger-in-charge
Mt Buffalo National Park
Mt Buffalo
Victoria 3745
Tel. (057) 55 1466

Children enjoy the snow at Mt Buffalo National Park.

An approaching storm over Mt Buffalo.

Great Ocean Road in Otway National Park provides magnificent views across Bass Strait.

Otway
National Park

Platforms of wave-worn sandstone skirting cliffs along the gale-ravaged Bass Strait coast, uninhabited southern ocean beaches, towering forests and picturesque fern gullies make Otway National Park a place of rare beauty and fascination for visitors and an ideal place for bush-walking, sightseeing, fishing and surfing.

Facilities

There are three main campsites within the park. The Aire River campsite offers toilets, a picnic area and canoeing; the Johanna campsite has toilets and a picnic area and is a good fishing location. The Blanket Bay campsite has barbecue areas, fireplaces, toilets and fresh water from the Parker River. The beach here is safe for swimming and snorkelling. Dogs on leads are permitted at the west of Johanna Beach and camping area and Princetown Beach. There is also a camping area at Princetown. A caravan park and hotel and motel accommodation is available at Port Campbell. Bush camping is not permitted in the park except for walkers. Accommodation is also available at Lavers Hill.

Location and Access

The park is located west of Apollo Bay, about 200 km south-west of Melbourne. Approach via the Great Ocean Road from Apollo Bay, Princetown or Lavers Hill.

Landscape

Otway National Park is an area of 12 750 ha, comprised of fine fern gullies, forest, woodland and heathlands as well as the spectacular coastline between Apollo Bay and Princetown which varies from coastal cliffs to sand dunes. Sandstone ridges border the cliffs along the shore east of Cape Otway and the eastern section of the park includes an inland wedge taking in the catchments of the Calder, Parker and Elliott Rivers in the Otway Range.

Flora and Fauna

Many different ecosystems, from sheltered mountain gullies to exposed dunes, support a great variety of plants. A high rainfall supports towering forests of mountain ash on protected slopes and gullies of myrtle beech rainforest with thick undergrowths of mosses and treeferns. Of particular interest are long clubmoss, skirted treefern, bog mallee and yellow leek-orchid.

Swamp wallabies and ring-tailed possums are commonly seen in the park, while the tiger quoll and yellow-bellied glider are present but rarely seen. Notable birds include the king parrot, powerful owl, rufous bristlebird, satin bowerbird and the white phase of the grey goshawk.

Walking Tracks

Several walking tracks and the beaches offer both short and extended walks. The sandstone platforms skirting the coastal cliffs along Bass Strait shore provide a pathway for walkers beginning a journey round the Cape and along the southern ocean beaches reaching towards Port Campbell. Given fair weather and timing of the tides, about 60 km can be traversed at or near the waterline, in four or five days.

Other Activities

The park offers opportunities for camping, walking, fishing, canoeing, surfing, sightseeing and nature study.

The Cape Otway lighthouse (adjacent to the park) can be visited. However, trips to the lighthouse must be prearranged. The grounds of the lighthouse are open Tuesday to Sunday inclusive from 10.00 a.m. until 4.00 p.m. Telephone (052) 37 9240 before you visit.

Climate

Spring to autumn are the best times to visit.

Bookings and Fees

No bookings or fees required.

Information

The Ranger
Otway National Park
P.O. Box 63
Apollo Bay
Victoria 3233
Tel. (052) 37 6889

The Ranger
Otway National Park
Post Office
Lavers Hill
Victoria 3238
Tel. (052) 37 3243

Port Campbell National Park

For millions of years the waves of the stormy Southern Ocean have pounded the limestone plateau around Port Campbell on Victoria's southern coast. The results may be seen in the spectacular gorges, archways, cliffs and blowholes which are featured in the Port Campbell National Park, 1750 ha of rugged scenic coastline between Peterborough and Princetown. The Twelve Apostles, London Bridge and Island Arch are some of the more renowned natural sculptures formed by the sea. Loch Ard Gorge is a steep narrow canyon named after the vessel Lord Ard, *one of the many shipwrecks which occurred in the seas around Port Campbell in the late nineteenth century.*

Facilities

A camping area in Port Campbell has powered and unpowered sites available for caravans and tents. Hot showers, toilets, laundry and barbecue facilities are also provided. Fires are not permitted so it is advisable to bring a portable fuel appliance. A visit to the information centre in Port Campbell would be helpful for advice concerning leisure activities, walking tracks and special attractions of the park.

Outside the park, accommodation is available in the motels and hotels in Port Campbell and the caravan parks at Peterborough and Princetown.

Location and Access

Port Campbell is situated on Victoria's western coast approximately 250 km from Melbourne. It is a comfortable three-hour drive by sealed road via Colac and Cobden. A slightly longer route along the Great Ocean Road which forms the landward boundary of the park offers rewarding views of the magnificent coast.

Toward the western end of Port Campbell National Park, London Bridge is rapidly being eroded by the wave power of the Southern Ocean.

The constant pressure of heavy swells and fierce storms on the limestone shoreline at Port Campbell National Park have created a series of ruggedly picturesque coastal formations.

Landscape

The park is a long narrow section of coastline and cliff tops through which Campbells Creek and the Sherbrook River flow into the sea. The onslaught of wind and waves have formed the limestone cliffs into fantastic patterns and shapes with the tougher remnants remaining as nearshore stacks and islands.

Sheltered beaches are found at the river mouths while the most striking sections of the coast can be explored from any point along the Great Ocean Road. Care should be taken near cliffs and lookouts.

Flora and Fauna

Coastal scrub and heathland, including such species as the silver banksia and dwarf she-oak, cloak the gorges and headlands of the park, forming a home for many different species of bird life. A small colony of fairy penguins survives near London Bridge, however the area has been closed to the public to avoid disturbance to the breeding areas. The park is important as a nesting place for the mutton bird. During summer these remarkable birds may be seen each night at dusk flying in to roost in the tussocky grass on Mutton Bird Island. Each year in April they depart on their annual 25 000 km migration to the north, around the Aleutian Islands and the shores of Alaska, returning in November to their Bass Strait rookeries.

Sheltered from the salt-laden coastal winds, vegetation towards the inland is taller and more dense. Scented paper-bark and the woolly tea-tree provide a suitable habitat for a variety of bird and animal life. Ringtail possums, wallabies and echidnas may be found while some of the common birds include the grey shrike-thrush, singing honeyeater and superb blue wren.

Walking Tracks

A self-guided 'Discovery Walk' starts near Port Campbell, taking about one-and-a-half hours. This provides an excellent introduction to the history and natural features of the park.

Access to the rock shelf near Port Campbell is obtained from Beacon Steps where walkers can experience the awesome power of the sea at close quarters. Care must be exercised here as waves do occasionally wash over the ledge.

Other Activities

While bushwalking and sightseeing are the most frequently enjoyed pastimes in Port Campbell National Park, water sports are also popular. Although the heavy surf makes swimming dangerous on the beaches, there are safe areas near the river mouth at the bay in Port Campbell. Fishing is recommended around Newfield Bay towards Peterborough and at Gibsons Steps near the Twelve Apostles, while canoeing is enjoyed at the mouth of the Gellibrand River at the eastern extremity of the park.

Previous page: *Sunset at the Twelve Apostles in Port Campbell National Park.*

Climate

During spring and summer the park is favoured with a temperate climate, however winter can bring cool conditions and chilling winds.

Bookings and Fees

Fees are payable for camping. Bookings are required for camping during Christmas holidays, Easter and Labour Day long weekend.

Information and Bookings

Port Campbell National Park
Tregea Street
Port Campbell
Victoria 3269
Tel. (055) 98 6382

The Grampians
National Park

A magnificent series of high, weathered ridges, sloping gently to the west and dropping in perpendicular cliffs to the eastern side form Victoria's most imposing new National Park. The brilliant spring wildflowers of the upland heaths, Aboriginal rock art and the many lakes, waterfalls and streams in Grampians National Park provide an outstanding setting for nature walks, picnics and camping.

Facilities

Public amenities such as camping grounds and picnic areas are widespread in the park. Caravan and campsites with toilets, showers and fireplaces are located at Zumstein on McKenzie Creek within the park and at Halls Gap, Wartook and Fyans Lake outside the park. Bush camping is permitted except in the water catchment areas and the Wonderland section. However, it is preferred that visitors use the established camping grounds where available.

Halls Gap also has several motels and a guesthouse, and nearby towns provide a variety of accommodation.

Location and Access

The park is located 260 km west of Melbourne and 25 km south-west of Stawell. It is reached by car off the Western Highway from Stawell. A train service operates from Melbourne to Stawell. Accessible roads traverse the park.

Landscape

The 167 000 ha Grampians National Park became Victoria's largest park in 1984 when most of the Grampians State Forest area was relinquished. The region, forming the westernmost heights of the Great Dividing Range, is comprised of folded and weathered sandstones, capped by Mt William (1168 m) in the east, with many summits exceeding 800 m. The park features numerous striking land forms including the Balconies near Halls Gap, the Broken Falls on the Mackenzie River, the Grand Canyon, a steep, narrow gorge deeply etched by rain and chemical action and the tough upthrust ridges of the Serra Range. Three sizeable lakes and countless creeks and waterfalls are included in the park estate.

Flora and Fauna

The heights carry scrubby vegetation and wildflower heaths where boronia, wattles, orchids and many other flowering plant species bloom in the spring. On the lower levels dry eucalypt forests and woodlands dominate but treeferns survive in the deep gullies as rainforest relics. The area is rich in wildlife; among the many mammals are kangaroo, koala, the smoky mouse found only in the Grampians, and platypuses in the streams. A number of freshwater fish inhabit the waterways and over 200 bird species have been recorded in the park.

Walking Tracks

The Wonderland section near Halls Gap has a number of nature walks with explanatory markings while hundreds of kilometres of tracks and trails are available elsewhere in the park.

Other Activities

Rock climbing, scenic drives, wildflower and bird observing, photography, boating and fishing are popular activities in Grampians National Park. A licence is required for fishing.

Bookings and Fees

There is no park entrance or camping fees, except at Zumstein where advance bookings can be made direct to the manager.

Information and Bookings
Ranger
C/o Dept. Conservation, Forests and Lands
Halls Gap
Victoria 3381
Tel. (053) 56 4381

Zumstein Camping Area
Via Horsham 3400
Victoria
Tel. (053) 83 6242

Left: *In late winter and early spring glorious golden wattle dots the Grampians' landscape.*

The Broken Falls on the Mackenzie River in The Grampians National Park.

The Lakes
National Park

Protected from the ocean by the high dune system of Ninety Mile Beach, the Gippsland Lakes are one of Australia's major inland waterways. Located on the south-eastern coast of Victoria, the lakes were once part of a huge bay, long since enclosed by a sand barrier transported by ocean currents. The Lakes National Park conserves the natural beauty and woodland of Sperm Whale Head and its lake shores, together with an abundance of wildlife.

Facilities
Car camping is permitted only at Emu Bay Bight where there are fireplaces, toilets, water and picnic tables. Picnic areas with fireplaces and toilets have been established at various locations around the lakeside; there are boat jetties at Point Wilson and Rotamah Island. The park headquarters is near the entrance in the western section of the park.

A caravan park, camping ground and holiday cabins are available at Loch Sport.

Location and Access
The park is 320 km east of Melbourne via the Princes Highway, turn off south at Sale to Longford and follow the sealed

Eastern grey kangaroos are a frequent sight at The Lakes National Park.

road to Loch Sport. Roads within the park are sandy. The park can also be reached by hired boat from Paynesville, 5 km away.

Landscape

The park covers 2380 ha comprising Rotamah Island and most of Sperm Whale Head. This peninsula is 14 km long and up to 2 km wide, and is separated by the shallow waters of Lake Reeve from the sand dunes of Ninety Mile Beach (part of the Gippsland Lakes Coastal Park). The low-lying land of the peninsula is covered with open woodlands and coastal heath with scattered freshwater swamps and saltmarshes. The artificial opening to the ocean made at Lakes Entrance in 1889 changed the lake waters from fresh to saline, significantly altering their ecology.

Flora and Fauna

Eastern grey kangaroos, swamp wallabies, possums, echidnas, wombats, koalas, bandicoots, native rats, antechinus and a herd of Asian hog deer inhabit the park. More than 200 bird species have been recorded, including emus, honeyeaters, red wattlebirds, scarlet robins, eastern rosellas, yellow-tailed black cockatoos and rainbow lorikeets. The lakes are home to swans, pelicans, herons, cormorants, ducks, royal spoonbills, sharp-tailed sandpipers and red-necked stints.

Eucalypt and banksia woodland covers most of the park. The main eucalypts are narrow-leaved peppermints, manna gums and forest red gums. Saw-leaf banksias are very common, with an understorey of bracken and such flowering plants as golden grevillea, common heath, silky tea-tree, fringe myrtle, silver banksia, wattles, ground orchids and pea flowers. In the freshwater swamp areas are swamp paperbarks, coast banksias, swamp she-oaks, prickly moses and coastal tea-trees.

Walking Tracks

A number of walking tracks and fire trails pass through the range of habitats on the peninsula and Rotamah Island. Many short walks originate in the Point Wilson area, and a nature trail begins at the lookout tower.

Other Activities

Canoeing, boating, fishing and swimming are popular activities. A number of birdwatching hides have been built in the park. Ninety Mile Beach offers good surfing and fishing.

Climate

The park has warm summers and cool winters with a low annual rainfall. Visit all year round.

Bookings and Fees

Campsites can be booked in advance, fees are payable and permits required from the ranger.

Information and Bookings

Ranger-in-charge
The Lakes National Park
Loch Sport
Victoria 3851
Tel. (051) 46 0278

The impressive inland waterway of the Gippsland Lakes.

A royal spoonbill at The Lakes National Park.

Wilsons Promontory
National Park

On the southernmost tip of the Australian mainland a great granite outcrop juts out into the windswept waters of the Bass Strait. The 'Prom', as Wilsons Promontory National Park is known to Victorians, is one of the state's largest and most popular National Parks. Packed into this relatively small finger of land are mountain peaks, sweeping white beaches, steep granite headlands, tall forests, fern gullies, saltmarshes and a wealth of wildlife.

Facilities
Caravan and car camping is permitted at Tidal River camping area where there are 500 unpowered sites, toilets, hot showers and laundry facilities. Open fires are banned in the south end of the park, and advance booking is essential during peak holiday periods. There are a number of camping areas for walkers with basic facilities in the southern section of the park, and bush camping areas without facilities in the northern section. The ranger should be consulted before bush camping.

Fully equipped flats and lodges accommodating two to six people can be rented. Group lodges with 12 to 24 beds are available, and there are economical motor huts with four beds.

There is a general store, take-away food shop, petrol station, post office, and an open-air cinema and a resident doctor in the summer. Many picnic areas are provided, with electric barbecues in the main car park at Norman Bay.

Location and Access
The park is 230 km south-east of Melbourne via the Gippsland Highway, turn off at Meeniyan or Foster. The main access roads are sealed.

Landscape
The park occupies 49 000 ha, and is mainly a granite mass forming rugged ranges rising to 754 m at Mt Latrobe. A sandy, densely vegetated isthmus joins the promontory to the mainland. There are grasslands, forested gullies, sand dunes and many beautiful sandy bays between rocky headlands. A number of creeks and rivers originate in the park.

Flora and Fauna
Eastern grey kangaroos, swamp and red-necked wallabies, koalas, wombats, possums and emus are often encountered.

Lorikeets, crimson rosellas, wattle birds and blue wrens are often seen around the camping area. Other common species include the grey shrike-thrush, eastern yellow robin, New Holland honeyeater and spotted pardalote.

Tall forests on the mountain slopes contain blue gums, messmates, yellow stringybarks and peppermint gums. There are pockets of rainforest, stands of lilly pillies and fern gullies.

The fairytale charm of rainforested Lilly Pilly Gully in Wilsons Promontory National Park.

Two views of Wilsons Promontory from Mt Oberon.

Splendid coastal views from Squeaky Beach Nature Walk in Wilsons Promontory National Park.

Along the sandy Yanakie Isthmus are big manna gums, coast banksias, drooping she-oaks, swamp paperbarks, coast tea-trees and coast wattles, with mangroves on the mudflats. Heathland areas with silky hakeas, dwarf she-oaks and many banksias burst into flower in the spring.

Walking Tracks
More than 80 km of walking tracks cross the park, varying from short walks of 1–6 km to long walks taking two days or more. Short tracks lead to beaches, bays and headlands, and many tracks lead from the Tidal River area. There are several self-guided nature walks, and in the summer and holiday periods the staff guide visitors on walks and spotlight tours.

Other Activities
Swimming, surfing, fishing and boating are all popular activities.

Climate
The area has warm summers and cool winters. Visit all year round.

Bookings and Fees
Christmas bookings for the Tidal River camping area are taken only in June, and a ballot is held to allocate sites. Bookings must be made for other holiday periods. Flats and lodges must be booked through the Victorian Government Travel Centre (tel. 03 602 9444). Group lodges and motor huts are booked through the park office. Fees are payable for all accommodation and park entry.

Information and Bookings
Ranger-in-charge
Wilsons Promontory National Park
Tidal River
Via Foster
Victoria 3960
Tel. (056) 80 8538

Wyperfeld National Park

In the north-western corner of Victoria, peaceful Wyperfeld National Park is one of the most fascinating parks in Australia. River red gums, native pines, mallee and heathlands shelter a wide range of dry country birds and animals. Lying on the final extremities of the Wimmera River system, Wyperfeld contains a series of usually dry lakes which only fill every few decades. During these times waterbirds arrive in great numbers.

Facilities

A large open space camping ground surrounded by towering red gums caters for most park visitors. Fireplaces, toilets and a picnic shelter are provided. Water is available but should be used sparingly. At the information centre visitors can obtain advice on special features of the park and current activities. Another camping area with basic facilities is located at the northern end of the park. There is also a small tranquil camping area accessible by car from Patchewollock.

For information on bush camping the ranger must be contacted. Water should be carried and no fires may be lit except in the fireplaces provided. Nearby Rainbow and Hopetoun offer a variety of other accommodation.

Location and Access

Wyperfeld is 450 km north-west of Melbourne and can be reached on a sealed road via Hopetoun or Rainbow. Access to the main camping and picnic area is from Yaapeet, the closest township which sells petrol and general supplies.

Landscape

Mallee vegetation is the dominant feature of Wyperfeld National Park. The landscape is flat and open, an extensive grey-green sea of mallee. Outlet Creek, the northern section of the Wimmera River, winds a tortuous course through the numerous sand dunes which connect the chain of shallow lake beds. These hold water only when the Wimmera River flows strongly enough into Lake Hindmarsh to the south, however Lake Albacutya to the north is occasionally filled. The creek and some of the lakes were flooded in 1918, 1957 and again in 1976 after which the creek filled for a year or two forming a silvery waterway through the woodland.

Sunset over Wyperfeld National Park.

Flora and Fauna

The characteristic vegetation at eastern Wyperfeld, known by the term 'mallee', refers to a peculiar form of adaptation common to some 20 species of inland eucalypts. The mallee trees are small and drought resistant with multiple stems sprouting from a large ligno tuber. One of the most common of these is the yellow mallee. Patches of scrub cypress pine, tea-trees, buloke and banksia are to be found on dunes near the lakes while in the spring the park boasts a colourful spread of wildflowers including some ground orchids, the poached egg daisy, golden pennants and guinea flowers. River red gum and black box woodlands cover the floodplains.

Emus and western grey kangaroos are often visible, feeding in the woodlands and on the dry lake beds. The bird life of Wyperfeld is outstanding and includes over 200 species. Sulphur-crested cockatoos, galahs and red-rumped parrots are plentiful. Smaller birds such as fairy wrens and red-capped robins may be found also.

The remarkable mallee fowl lays its eggs in specially prepared mounds of rotting vegetation and the male is responsible for keeping the mound at a special incubation temperature before the chicks hatch.

Walking Tracks

Tyakil and Lake Brambruck nature walks are self-guided tracks which give visitors a close-up look at native flora and fauna in Wyperfeld. These begin near the main camping ground and leaflets are available at the start of the tracks.

Other Activities

Bushwalkers and naturalists can enjoy an abundance of animal and plant life. The Eastern Lookout Nature Drive is recommended as a quicker means of viewing the features of the park. Cycling is popular along the Eastern Lookout and Entrance Roads as well as the Outlet Creek track.

The grey-green sea of mallee scrub is the dominant ground cover at Wyperfeld.

Climate

Summer is extremely hot so the best time to visit the park is in autumn, winter or spring.

Bookings and Fees

Fees are charged for camping but bookings are not required.

Information

Ranger-in-charge
Wyperfeld National Park
P.M.B. 1465
Yaapeet
Victoria 3424
Tel. (053) 95 7221

The sulphur-crested cockatoo. These birds gather around the waterholes of Wyperfeld in large noisy flocks.

Left: *Exposed slopes in Wyperfeld National Park support a dense growth of wind-blown banksia woodland.*

South Australia

Pelican flotillas, wild surf beaches, rippling sand dunes, swamplands, high forested ranges and the deep gorges of the Flinders are the scenic highlights of South Australia's National Parks.

Nowhere else in Australia can wildlife be seen in such close proximity and in such profusion as in the parks of this state. To protect its valuable native animals and birds in their variety of habitats and to conserve the natural features of the landscape, South Australia has set aside five per cent of its total area as national conservation and recreation parks.

Climate bears importantly on a choice of destination — the coastal parks are cool in summer and autumn, while mountain areas are ideal to visit in winter and spring. The coastal parks include Lincoln National Park with its popular attractions of seaside pastimes; the boisterous surf and gentle landlocked waters of Coffin Bay National Park; and one of the most magnificent National Parks in Australia, the Coorong, which is a series of saltwater lagoons fed by the waters of the Murray River. One of the last natural bird sanctuaries in Australia, giant pelicans, ibis, shags, swans and terns are all at home here, their piercing cries combining with the thundering Southern Ocean.

Flinders Chase National Park, on Kangaroo Island, is the state's most important wildlife refuge. Here, the picnic grounds are fenced, not to hold the people in but to keep the friendly native animals out.

Inland, on the southern edge of the Flinders Ranges is Mt Remarkable National Park, an area of cool, deep gorges, tree-dotted valleys and grasslands where wildflowers bloom brilliantly in early spring. North of Adelaide, the Flinders Ranges National Park covers a great tract of rough ranges cut by deep gorges. In few places does their sparse vegetation cover the reds and browns of rock and earth. This sculptured and eroded landscape reaches a scenic climax at Wilpena Pound, a huge basin encircled by outward facing cliffs.

Coorong
National Park

South of the mouth of the Murray River the thin strip of sand known as the Younghusband Peninsula stretches far into the misty haze of the horizon. The great rollers of the Southern Ocean ceaselessly pound their rhythm on the sand, sculptured by the wind into Saharan-like ridges and valleys. Hundreds of pelicans wheel and glide overhead before coming to rest and feed on the waters of the Coorong, a long, narrow lagoon between the peninsula and the mainland. One of Australia's natural wonders, this stretch of water and beach, paralleling the coast for 145 km, is enclosed in the Coorong National Park, one of the most magnificent parks in South Australia.

Facilities
Camping is permitted anywhere in the park, although the only facilities available are water and toilets at the park headquarters at Salt Creek. There are privately run cabins, and caravan and camping grounds just outside the park at Salt Creek and Policeman's Point with all facilities, supplies and petrol. A second ranger station is located at Noonameena. Firewood is scarce and visitors should bring their own cooking fuel.

Location and Access
The park begins about 180 km south-east of Adelaide and extends for 145 km along the coast. Access is via the sealed Princes Highway, turn off along the unsealed road to Noonameena for the northern entrance. The beach is accessible to four-wheel-drive vehicles at Tea Tree and 42-Mile Crossing. Tea Tree Crossing is often under water in winter, and vehicles must stay on marked tracks and on hard, damp sand to prevent damage to the environment.

Landscape
The park covers 38 984 ha, including the long, shallow, saline lagoon, separated from the ocean by the narrow peninsula of massive sand dunes, vegetated in many areas. A number of islands in the lagoon are breeding grounds for birds, and visitors must not approach nearer than 140 m. Some of the northern lagoon islands are Aboriginal Reserves, and should not be visited.

Flora and Fauna
Western grey kangaroos are often seen on the mainland, but the main attraction is the prolific coastal bird life. This includes grey teal, chestnut teal, mountain ducks, black swans, spoonbills, ibises, cormorants and grebes. Australian pelicans, terns and seagulls have established large breeding colonies on the islands where up to 4000 pelicans can be seen in good seasons. Many land birds occur in the dunes, including the orange-bellied parrot, singing honeyeater and bronzewing pigeon.

Seagulls feeding on the shores of the Southern Ocean at sunset in Coorong National Park.

A pelican preening in Coorong National Park.

Spinifex, sea spurge, marram grass and two-horned sea rockets bind the sand of the foredunes, with coastal wattles, mallees and sandhill daisies on the consolidated dunes. Teatrees, samphires, swampy paperbarks, banksias and grass trees surround the lagoons.

Walking Tracks
Tracks lead from the road to the beach and lagoons, and a nature trail explores the lakes in the Salt Creek area.

Other Activities
The beach is popular for fishing and surfing. Boating on the lagoon is an excellent way to view the bird life, but care should be taken for reefs and sandbars. Shooting of some waterfowl is permitted in the Coorong Game Reserve near Salt Creek for a limited season each year.

Sunrise over the Younghusband Peninsula at Coorong National Park.

Climate
The park has warm summers and cool, wet winters. Visit in spring, summer and autumn.

Bookings and Fees
Camping permits must be obtained from the ranger.

Information
Ranger-in-charge
Coorong National Park
P.O. Box 105
Meningie
South Australia 5264
Tel. (085) 75 7014

Dry and wind-blown coastal scrub on the Younghusband Peninsula.

Flinders Chase
National Park

Kangaroo Island is Australia's second largest island. Its isolation from the mainland, its small population and slow development, have conspired to create a sanctuary for a wealth of native flora and fauna. Flinders Chase National Park comprises almost the entire western part of the island, an area rich in natural features and native animals completely at ease with their human visitors. Here are colonies of seals playing in the thundering waves, platypuses, koalas and the ubiquitous kangaroos.

Facilities
The main caravan and car camping area is near the park headquarters at Rocky River, where there are gas barbecues, picnic tables, toilets and hot showers. A number of bush camping sites and picnic areas have been established in the park. Cottages are available for hire by small groups at Rocky River with beds, electricity, water and fireplaces.

Accommodation is also available at Kingscote and American River and Hanson Bay about 9 km east of Flinders Chase.

Location and Access
The park is 97 km from Kingscote via the sealed Playford Highway, turn off along the unsealed West End Highway into the park; or via the unsealed South Coast Road. Access from the mainland is via the car ferry from Port Adelaide to Kingscote (seven hours), from Cape Jervis to Penneshaw on a fast passenger ferry, or on a 30-minute flight from Adelaide. Roads within the park are unsealed.

Landscape
The park occupies 73 103 ha of heavily vegetated low sand hills in the south and forested plateau in the north. High cliffs, broken by sandy bays and estuaries form the coastline. At Kirkpatrick Point in the south-eastern corner huge granite boulders sculptured by the elements dominate the area. Below the cliffs at the south-western Cape du Couedic, fur seals congregate in the stalactite-fringed cave known as Admiral's Arch.

Flora and Fauna
The most commonly seen mammals are the Kangaroo Island kangaroo, tammar wallaby, echidna, brush-tailed possum, koala, Australian sea lion, the New Zealand fur seal and,

Top right: *The old lighthouse-keeper's cottage with Cape du Couedic Lighthouse in the background.*

Centre right: *Australian fur seals at Seal Bay on Kangaroo Island.*

Bottom right: *The stalactite-fringed cave of Admiral's Arch below the cliffs at Cape du Couedic on Kangaroo Island.*

OCEANOGRAPHICAL
FIELD STATION
Nº2

The Remarkable Rocks, richly coloured granite boulders sculptured by the wind and the ocean on the southern shore of Kangaroo Island.

occasionally, the short-nosed bandicoot. The platypus and ringtail possum have been introduced. Gould's sand goanna, black tiger snakes and copperheads may be seen.

Some 150 bird species inhabit the park. Yellow-tailed black cockatoos, Cape Barren geese, scarlet robins and honeyeaters are plentiful. Around the coast are silver gulls, ospreys, white-bellied sea eagles, little penguins and albatrosses.

Open woodlands of stringybark and mallee cover the plateau area north of Rocky River. Eucalypt forests follow the streams with tall stands of sugar, pink and swamp gums interspersed with banksias, tea-trees and sedge swamp areas. Dense mallee scrub covers the southern area, with stunted heathlands on the rocky cliff tops.

Walking Tracks

A number of walking tracks lead from the road to the coast, varying from a short nature trail to long walks along the cliff tops, to natural caverns, along river banks, through forests and scrublands. Much of the central area of the park is a trackless wilderness accessible only to experienced bushwalkers.

Other Activities

There are good fishing spots on the coast, however anglers should exercise caution when rock fishing as freak waves breaking without warning are not uncommon. Many opportunities exist for birdwatching, nature study and photography. Visit nearby Seal Bay where a large seal colony uses the wide beach to breed and rear their young in spring and summer.

Climate

The park has warm summers and cool, wet winters. Visit in spring, summer and autumn.

Bookings and Fees

Campsites cannot be booked, although permits must be obtained from the ranger and fees payed. Cottages must be booked through Ms B. Bell after business hours (tel. 0848 35 298).

Information and Bookings

Ranger-in-charge
Flinders Chase District Office
National Parks and Wildlife Service
P.M.B. 246
Via Kingscote
South Australia 5223
Tel. (0848) 37 235

Right: *Delightful purple nightshade flowers in the Flinders Ranges.*

Flinders Ranges
National Park

The sculptured and eroded mountains of the Flinders Ranges are one of the most ancient landscapes on earth, formed more than 500 million years ago and raised from the seabed by the enormous forces of the cooling planet. Flinders Ranges National Park incorporates much of the central Flinders Ranges. Its gateway to the south is in the mighty rock face of Wilpena Pound, rising more than 1000 m from the surrounding dry plains, opening up vistas of wild and rugged ranges, abundant wildlife, a wealth of wildflowers and some of the finest bushwalking in Australia.

Facilities
The main caravan and car camping area is at Wilpena, near the Wilpena Pound Motel. The area has fireplaces, toilets, hot showers, a laundry, store, petrol supplies and a ranger station. Other car camping sites with fireplaces, toilets and water are located at Oraparinna near the park headquarters, at Enorama Creek, The Guardian and Aroona Ruins.

A number of other camping areas without facilities have been established. Bush camping is permitted, but the ranger must be consulted about suitable sites. Firewood is scarce, and visitors should bring their own cooking fuel.

Location and Access
The park is some 350 km north of Adelaide. Access is via the sealed road from Hawker, 35 km south, or via the unsealed road from Blinman, 13 km north. Roads within the park are gravelled and may become impassable after heavy rain.

Landscape
The park covers 80 265 ha of rugged mountains, rising to 1165 m at St Mary's Peak to the north of Wilpena Pound: a vast, oval amphitheatre forming the southern section of the park. The Heysen Range forms the eastern boundary, cut through by creeks that have created the deep and spectacular Brachina Gorge, and the Bunyeroo and Edeowie Gorges. Rugged hills give way to grassy plains in the east, rising to form The Bunkers in the north-eastern corner.

The solitary Casneaux tree stands on a grassy plain in the east of the Flinders Ranges.

Flora and Fauna
Red kangaroos and euros are common, and yellow-footed rock wallabies may sometimes be seen. Other common mammals include the western little mastiff-cat, bats, foxes, goats and rabbits.

More than 100 bird species have been recorded in the park. They include emus, wedge-tailed eagles, brown falcons, the Australian kestrel, flocks of galahs, honeyeaters, thornbills, grasswrens and fairy wrens, magpies and ravens.

Previous page left: *Point Bonney in the Flinders Ranges.*

Previous page right: *The grass plains and ancient sculptured landscape of Flinders Ranges National Park.*

Towering red river gums line the creeks in this dry country, while groves of callitris and stands of sugar gums break up the grassy plains where wildflowers proliferate in springtime. Eucalypts and native pines grow on the slopes of the ranges.

Walking Tracks

There are many walking tracks leading to spectacular mountain top lookouts and beautiful gorges. Constructed trails lead into Wilpena Pound from the campsite, varying from two to eight hours in duration. Others follow creeks and ridges.

The Ranges offer many challenges and rewards for bushwalkers, but the rugged terrain is very demanding. Sufficient drinking water must be carried. Bushwalking guidelines and information are available from the park offices.

Other Activities

The rocky terrain offers many opportunities for rock climbing (inform the ranger beforehand). The park is a favourite area for landscape artists and photographers. Scenic flights can be made from the park to the surrounding lakes and mountains.

Climate

The area has hot, dry summers and cool winters. Visit in autumn, winter and spring.

Bookings and Fees

Camping permits must be obtained from the ranger. No bookings are accepted, and camping fees are payable at the Wilpena Pound camping area.

Information

Ranger-in-charge
Flinders Ranges National Park
Oraparinna
South Australia 5730
Tel. (086) 48 0017

Lincoln
National Park

Connected to the Eyre Peninsula by a narrow land bridge, Lincoln National Park is almost surrounded by ocean, and boasts some of the best coastal scenery in South Australia. This hilly headland on the south-eastern tip of Eyre Peninsula presents a wall of precipitous cliffs to keep back the forces of the wild Southern Ocean in its western and southern extremities, while a host of small islands break up the ocean swell in the east. The wide sandy bays and protected coves shaded by tall mallees offer excellent camping areas in the northern tip of the park, a perfect place to observe the magnificent and colourful bird life of the area.

Facilities
Caravan and car camping areas are located at Taylor's Landing and Surfleet Cove where there are fireplaces and toilets, and a boat ramp at Taylor's Landing. There are no powered sites and no showers. Caravans are also allowed at Stamford Beach campsite, although there are no facilities, and a number of car camping areas have been established on the northern tip of the park. The camping area at Memory Cove is only accessible by boat and four-wheel-drive vehicles, and water must be carried. Camping outside designated areas is not permitted.

Cape Donnington Cottage is available for school and community groups. Accommodation is available at Port Lincoln.

Location and Access
The park is 25 km south of Port Lincoln via the sealed road to Tulka, turn off here on the unsealed road to the park. Roads within the park are unsealed, and may be difficult for caravans. Check road conditions at Port Lincoln. MacLaren Point and Memory Cove are not accessible by conventional vehicles.

Landscape
The park occupies 17 085 ha of low country at the eastern tip of the Eyre Peninsula. It consists of a rugged headland surrounded by many small islands. A low range of granite hills and cliffs form the eastern side of the park, while precipitous cliffs protect the southern coast, and sand dunes bear the brunt of the southern gales in the south-west. Low limestone cliffs along the northern coast separate wide beaches and sheltered coves.

Flora and Fauna
Western grey kangaroos and bush rats are common, while dolphins and whales are sometimes spotted offshore. Sand monitors, bearded dragons, jacky lizards and peninsula brown snakes live in the mallee, and the peninsula dragon can be seen basking on rocky outcrops.

River red gums dot the valley floors of Mambray and Alligator Creeks.

The ferocious looking bearded dragon, one of the many reptile species seen in Lincoln National Park.

The brilliant plumage of the Port Lincoln parrot can often be spotted in the mallee, together with common bronzewings and tawny-crowned honeyeaters. Other common birds include the southern emu wren, white-browed scrub wren, western calamanthus, grey butcher birds, scarlet robins, western shrike-thrushes and golden whistlers. Along the coast are oyster catchers, dotterels, greenshanks, sandpipers, white-breasted sea eagles and ospreys.

Mallees dominate the vegetation with an understorey of native cherry, wattles, guinea flowers and velvet bush. Bands of coastal heath separate the mallee from the cliffs, with occasional stands of she-oaks and tea-trees on the limestone ridges.

Walking Tracks
A network of unmarked walking tracks follow the coastline, leading to spectacular lookouts and sheltered bays. Experienced bushwalkers can hike into the wilderness areas in the southern section of the park, but sufficient food and water must be carried.

Other Activities
The park is a haven for birdwatchers, there are good fishing spots and many areas to explore by boat. Swimmers are advised to be cautious as there many sharks in the area. Aboriginal campsites and stone artefacts are conserved within the park.

Climate
The park has warm summers and cool, wet winters. Visit in spring, summer and autumn.

Bookings and Fees
Bookings are not accepted for the campsites, but permits must be obtained from the ranger and fees paid. Cape Donnington Cottage must be booked in advance.

Information
District Ranger Eyre District Office
National Parks and Wildlife Service
P.O. Box 866
Port Lincoln
South Australia 5606
Tel. (086) 82 3936

Mount Remarkable
National Park

In a rugged, densely wooded tract of the southern Flinders Ranges, Mount Remarkable National Park incorporates the former Mambray Creek and Alligator Gorge parks where quietly flowing streams pass through deep and narrow tree-lined gorges of red-brown sandstone.

Facilities
There are picnic sites with tables and fireplaces at Alligator Gorge and Mambray Creek and camping sites with water, fireplaces and toilets. Bush camping is generally only allowed in designated sites but may be permitted elsewhere at the ranger's discretion. From November to April a total fire ban is enforced and no bush camping is permitted.

Accommodation of various kinds is available at nearby Port Pirie or Port Augusta. Caravan parks are available at Port Germein, Wilmington and Melrose.

Location and Access
The park is situated 270 km north of Adelaide between Port Pirie and Port Augusta. Access to Mambray Creek is via Highway 1, turning off 45 km north of Port Pirie. Alligator Gorge is reached via Main North Road and Wilmington.

Landscape
Mount Remarkable (995 m) is the highest point in the 8600 ha park, capping a densely vegetated range which is chiefly responsible for the relatively reliable rainfall of the immediate district. Mambray and Alligator Creeks, dissecting tilted outcrops of red-brown quartzite, flow through a large fertile oval basin rimmed with sharp ridges where sheep grazed until the 1960s. Regeneration of vegetation since grazing has been considerable.

Left: *High cliffs of the Battery Ridge along the Hidden Gorge trail in Mount Remarkable National Park.*

Flora and Fauna

The basin valleys are dotted with river red gums, white cypress pine and grasslands where brilliant wildflowers such as brush pea and native hibiscus bloom in springtime. Sugar gums and blue gums cover the ridges providing a home for marsupials including euros. Among the many reptile species is the big lace monitor. Rocky slopes in selected areas provide safe habitat for yellow-footed rock wallabies which are an endangered species. Abundant bird life in the park includes brown falcons, little eagles, black-shouldered kites, ringneck parrots, wedge-tailed eagles, painted quail and rainbow bee-eaters.

Walking Tracks

There are a number of excellent well-marked walking tracks around Mambray Creek and Alligator Gorge.

Other Activities

Bushwalking, birdwatching, camping and sightseeing are the chief activities enjoyed here.

Climate

The best times to visit the park are in spring, winter and autumn as summer is very hot and dry, and fire danger can be extreme.

Bookings and Fees

A permit is required for camping for which a fee is charged. This is available from the ranger and bookings are recommended for the Easter period and October Labour Day long weekend.

The weathered red cliffs of Alligator Gorge in Mount Remarkable National Park.

Information and Bookings

Ranger-in-charge
Mt Remarkable National Park
Private Bag 7 Port Germein
South Australia 5495
Tel. Mambray Creek (086) 34 7068
Alligator Gorge (086) 67 5181

The large prehistoric lace monitor can be seen at Mount Remarkable National Park.

Northern Territory

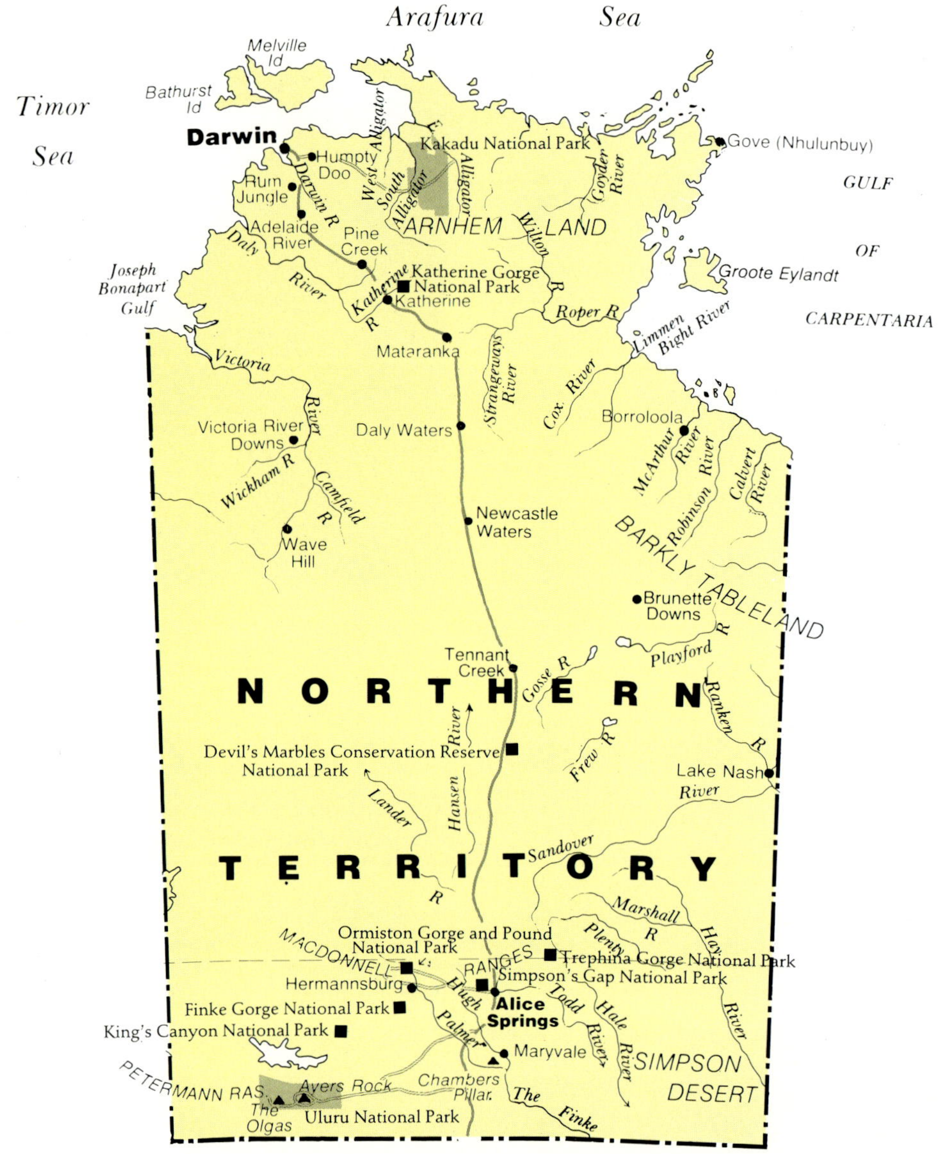

A land of surprising contrasts, the Northern Territory contains rugged ranges of bare rock and harsh spinifex, shadowy canyons with deep pools of cold water, great sand dunes of graceful curves, deep lagoons, crocodile infested rivers, snaking zig-zag gorges and towering cliffs.

Rising abruptly in stark, jagged crests, the great curve of the MacDonnell Ranges makes a 400 km wall across the very heart of Australia. Best known of all the National Parks in this area is Uluru, where, in their awesome grandeur, stand Ayers Rock and the Olgas.

Within the West MacDonnells are Kings Canyon National Park, containing a maze of gorges occupied by seasonal watercourses, and Finke Gorge National Park, a wilderness area that straddles the Finke River and includes the picturesque Palm Valley, a refuge for cycad palms. Here, too, is Ormiston Gorge and Pound National Park, where fish bury themselves as a string of waterholes shrink to puddles, waiting for the rains to brim them again.

Closer to Alice Springs are found the broken ridges of the Chewings Range in Simpsons Gap National Park and the nature park of Trephina Gorge.

Further north, huge tors of granite strewn among hummock grasses comprise the startling landmarks known as the Devils Marbles.

In the far north of the Northern Territory, the remorseless regime of monsoon and drought has created a unique environment. Here the impressive Katherine Gorge National Park is to be found, where anglers will be delighted with good catches, wildlife abounds and on the gorge walls, high above flood levels, Aboriginal rock paintings form large murals.

The other National Park in the Top End of the Territory is Kakadu, which contains two types of wilderness area: the blacksoil floodplains and paperbark lagoons of the Alligator River system and the rugged walls of the Arnhem Land escarpment which is deeply indented with streams, gorges and waterfalls.

Devils Marbles

Conservation Reserve

Gigantic rounded granite boulders, some balanced precariously upon one another and others strewn across the hummock grasses as if scattered carelessly by a giant hand, present a startling landmark on a bare and largely featureless plain in the central Northern Territory. The Marbles are impressive photographic objects particularly at sunset when they glow with a deep red colour.

Facilities

Picnic grounds with fireplaces and toilets are available in the reserve. Camping is permitted but fires should only be lit in barbecues provided.

The nearest settlement is at Wauchope, 9 km to the south, where petrol and stores may be obtained. Camping grounds and other accommodation are available at Tennant Creek, 97 km to the north of the reserve.

Location and Access

The reserve is located 393 km north of Alice Springs on both sides of the old Stuart Highway, a sealed road. There is road access through the park and visitors are asked to drive only on these.

Landscape

The Devils Marbles Conservation Reserve, covering 1800 ha, is a wide and shallow quartz plain with residual granite tors and low, bare granite outcrops. The original granite masses from which the Marbles were fashioned have eroded along the joints and, together with the flaking away of thin slabs from the surface, the corners have been gradually rounded into spherical shapes. Where the surrounding granite has been completely worn away, the boulders have been left one upon another in great heaps.

Flora and Fauna

The shade and dew run-off from the boulders creates a microclimate of miniature refuges and cooler, more sheltered environments in which small plants flourish and birds abound. Fairy martins construct bottle-shaped mud nests on the underside of overhanging boulders. The large sand goanna is often found among the thick spinifex growing around

Ghost gums growing in the shade of the Devils Marbles.

The huge rounded granite boulders of the Devils Marbles are known as the Eggs of the Rainbow Serpent in Aboriginal dreamtime myths.

the boulders while the pygmy spiny-tailed goanna is occasionally seen in the crevices. Zebra and painted finches may be found through the reserve. Some ghost gums flourish in the moist fissures of the Marbles.

Other animals include those generally encountered in arid regions such as wallabies and euros. Away from the boulders, very sparse mulga scrub, scattered low ground cover and occasional small gums grow. The peculiar caustic vine, sometimes found growing amongst the spinifex, has grey cactus-like branches with tiny yellow waxen flowers.

Walking Tracks

There are no marked trails through the reserve but the landscape provides easy walking country. Overnight campers should carry water and a compass.

Other Activities

Sightseeing and photography are the most popular pursuits at the reserve.

Climate

Late autumn through to early spring is the best time to visit the park as the rest of the year is subject to extreme heat conditions.

Bookings and Fees

No bookings or fees are required.

Information and Bookings

Ranger
Devils Marbles Conservation Reserve
C/o Northern Territory Conservation Commission
Gap Road
Alice Springs
Northern Territory 5750
Tel. (089) 50 8211

Northern Territory Conservation Commission
P.O. Box 394
Tennant Creek
Northern Territory 5760
Tel. (089) 62 2140

Finke Gorge National Park

At Finke Gorge National Park the meandering Finke River has cut deeply into the surrounding ranges twisting amongst 25 km of arresting sandstone rock formations. The prehistoric palms and tranquil rock pools of Palm Valley are a focal point of this wilderness area which covers over 45 000 ha west of Alice Springs.

Facilities

A picnic area and a camping ground with gas barbecues, toilets and showers are provided within the park and there is a ranger in residence. Bush camping is permitted but the ranger must be consulted first and firewood should be collected from outside the park.

Adequate food, water and fuel supplies should be carried and these are available at Hermannsburg, 16 km to the north of the park entrance.

Location and Access

Finke Gorge National Park is situated 138 km west of Alice Springs. Most of the road is unsealed and for the last 16 km a four-wheel-drive vehicle is required. After rains this section may be temporarily impassable. Daily coach tours operate from Alice Springs.

Landscape

Much of the southern portion of the park consists of a sandplain wilderness but towards the north impressive red sandstone formations have been carved by wind erosion and the watercourses. Near the junction of Palm Creek and the Finke River, both usually dry and sandy with strings of waterholes, Kalarranga Lookout and a striking ridge-ringed basin known as the Amphitheatre are featured.

Upstream, Palm Valley is a wide canyon where tall palms sheltered beneath the high cliffs grow prolifically beside temporary rockpools.

Cabbage palms in Palm Valley at Finke Gorge National Park are visible evidence of a time when seas invaded Central Australia.

The pied butcher bird found at Finke Gorge National Park.

Flora and Fauna

The park was established to protect the comprehensive collection of living central Australian plants established there, especially the cycads and cabbage palms growing in the damp gorges, relics of a much wetter time. Ghost gums, ferns and mosses also grow along the gorges contrasting with the bare cliffs and barren ranges beyond. Even here though, eucalypts, drought resistant shrubs and spinifex grow, together with a variety of spring-blooming wildflowers such as the silver-leaf cassia and the *Acetosa* or wild hop.

Abundant bird life can be seen including cockatoos, wedge-tailed eagles, the black-faced cuckoo-shrike and the musical pied butcher bird. Many reptiles and a variety of freshwater fish are to be found while mammals include kangaroos, wallabies, euros and dingos.

Walking Tracks

There are short walking trails leading from the camping ground to Palm Bend and the Amphitheatre. Information on longer walks is available from the ranger.

Other Activities

Photography is especially popular and swimming in Palm Creek may be enjoyed when there is sufficient water.

Climate

Summer is very hot, so from autumn to spring is the most pleasant time to visit the park.

Bookings and Fees

There are no entrance fees or camping charges. Campsites are allocated on a first come first serve basis.

Information

Ranger
Finke Gorge National Park
C/o Northern Territory Conservation Commission
Gap Road
Alice Springs
Northern Territory 5750
Tel. (089) 56 7401

Kakadu
National Park

In the remote tropical north of Australia, where the Alligator Rivers pour from the cracked and craggy Arnhem Land escarpment to wind their way through vast floodplains to the great tidal estuaries of the Van Diemen Gulf, live the survivors of a culture born 40 000 years ago. Part of this fascinating area has been set aside as the Kakadu National Park and recorded on the World Heritage List as a place of outstanding cultural interest and natural beauty. Here is the most outstanding concentration of Aboriginal art in Australia, and one of the country's most magnificent natural reserves.

Facilities

There are eleven camping areas in the park, five accept caravans, and all have toilets and fireplaces, however powered sites are not available. Muirella Park, East Alligator and Mardugal camping areas are the most developed with showers and other facilities, and are very popular during school holidays and peak visitor times. Four sites are now open throughout the wet season. Bush camping is allowed, but a permit must be obtained from the ranger.

Beware of crocodiles. If camping near a billabong or creek do not leave food or refuse lying around, and keep children away from the water's edge. Do not swim or go in creeks, rivers or billabongs.

Motel-type accommodation is available at the South Alligator Motor Inn and Jim Jam Hotel at Cooinda inside the park. Food and petrol are available at the Mel camping area, Jabiru, Cooinda and the South Alligator Motor Inn. Drinking water should be carried and billabong water boiled before drinking.

Dramatic Aboriginal cave paintings at Nourlangie Rock in Kakadu National Park.

There are a number of picnic areas and boat launching sites at the East Alligator, South Alligator Inn, Yellow Water and Mardugal areas. The visitor centre is near Jabiru township.

In the tradition of thousands of years of Aboriginal habitation of the Kakadu region, Aboriginal children play in the East Alligator River.

Location and Access

The park is 250 km east of Darwin via the sealed Arnhem Highway. From the south access is via the largely unsealed road from Pine Creek (206 km) which is often closed in the wet season. The main access road within the park is sealed, others are unsealed and many are impassable during the wet season. Contact the park headquarters for up-to-date road information.

Landscape

The park occupies 568 400 ha of mainly low-lying land dominated by the floodplains of the East, South and West Alligator Rivers. The red walls of the Arnhem Land escarpment rise abruptly from the undulating plains in the south-eastern section of the park. The sandstone plateau at 250 m high is cracked and rugged, and gives rise to the great river systems of the area which pour from the rocks in cascades and rapids.

On the lowlands are scattered rainforest areas and large woodlands. The river systems expand into wide lagoons in the wet season, forming a vast floodplain dotted with paperbark trees. Mangrove forests line the banks of the wide river estuaries which turn into great mudflats at low tide.

Flora and Fauna

Euros, short-eared rock wallabies, black wallaroos, rock possums, northern quolls and bats live among the rocky slopes of the plateau. Marsupial mice, pygmy and ringtail possums, dingos and destructive feral pigs and water buffalo are occasionally seen in the lowlands. Saltwater crocodiles are common in both fresh and salt water, and make it extremely dangerous to swim in the area. Other reptiles include goannas, water pythons, keelback snakes, brown and green tree snakes, Mitchell's water monitor and the large frilled lizard. Barramundi and many other fish species are plentiful.

Bird life is prolific. More than one-third of the Australian species have been seen in the park, together with many migratory visitors. Waterbirds often congregate in their thousands. There are pelicans, geese, ducks, herons, egrets and the comb-crested jacana which walks delicately on the water-lilies. The woodlands contain parrots, cockatoos, lorikeets, hawks, pigeons, quails, owls, honeyeaters, friarbirds and finches. Birds unique to this area include the chestnut-quilled rock pigeon, the banded pigeon, white-lined honeyeater and white-throated grasswren.

Mangrove forests occur on the salty tidal flats, with acacia, pandanus and eugenia species on the beach ridges. Paperbark forests characterise the floodplains, with sedges, grasses and *Sesbania* on the wetlands. The lowlands support eucalypt woodlands, where *E. miniata* and *E. tetrodonta* are the most common species, and scrubby areas with pandanus, acacia,

A blue-winged kookaburra in Kakadu National Park.

Water buffalo in Kakadu National Park.

Fishing at Yellow Water which drains into the South Alligator River.

grevillea, melaleuca, grasses and sedges. Monsoon forests occur between the lowlands and the floodplains, while on the plateau are isolated pockets of eucalypt woodland and rainforest among spinifex and wiry grasses.

Walking Tracks

A number of well-defined walking tracks have been established from nature trails, some leading to Aboriginal galleries, to long hikes along creeks, through woodlands and rainforests to caves, billabongs and the rocky escarpment. Experienced bushwalkers will find many challenging and interesting areas to explore, but the ranger should be consulted beforehand, and drinking water carried.

Other Activities

This is one of the richest areas in Australia for Aboriginal paintings. Ubirr, with its vivid 'X-ray' studies, and Nourlangie Rock, where there are dramatic series of human and god-like figures, contain excellent examples of the different styles. Fishing is a popular activity, regulations apply, and the ranger should be consulted. Boating on the East and South Alligator Rivers is an excellent way to see the landscape and wildlife, but beware of fast-flowing water, tidal changes and hidden snags. Organised boat tours are conducted by private operators.

Pied herons, one of the numerous species of waterfowl found in Kakadu National Park.

Left: *Saltwater crocodiles, natural inhabitants of the Kakadu wetlands, are often seen basking in the sunshine along the river banks.*

Climate
Kakadu has two distinct seasons: a wet, hot season from November to March, and a dry, cooler season for the rest of the year. June to August is the best time to visit.

Bookings and Fees
Bookings are not accepted for the campsites, and visitors should check availability of sites during peak holiday periods. Camping fees are payable at the privately run grounds but not at park-operated areas.

Information
The Superintendent
Kakadu National Park
P.O. Box 71
Jabiru
Northern Territory 5796
Tel. (089) 79 2101

Katherine Gorge
National Park

A mighty gorge cut through a rock plateau by the torrential annual flooding of the Katherine River is the focal point for the massive 180 000 ha Katherine Gorge National Park. Ancient sandstone walls tower above the waterway where every dry season thousands of visitors enjoy boat tours through this spectacular scenic region.

Facilities
At the entrance to the gorge a tourist camp and caravan park is privately run under the supervision of the park ranger. Powered caravan sites, barbecues, toilets, showers and laundry facilities are provided. Picnic areas with barbecues and toilets and a visitor centre are also located nearby.

Bush camping is permitted along the river but camping permits must be obtained from the visitor centre. Launch tours also operate daily throughout the dry season and there is a boat ramp for private vessels.

Location and Access
The park is situated 330 km from Darwin via the Stuart Highway and a turnoff at Katherine travelling north-east for 32 km. Edith Falls in the north-west of the park may be reached by sealed road turning off 40 km north of Katherine, but this road is sometimes closed in the wet season.

Landscape
The canyon through which the Katherine River meanders is cut in many places by the narrow chasms of tributary creeks bordered by steep sandstone and conglomerate cliffs. Upstream the gorge widens into a broader valley which forms the southernmost portion of the Arnhem Land plateau.

During the wet season brown floodwaters surge through the gorge, but in the dry the river flows quietly in long deep reaches between the rapids. Aboriginal paintings on the cliff walls depict native animals several times their lifesize. On the plateau amidst the dry sandy woodlands are intermittent areas of swampland.

Flora and Fauna
The river banks and gorges are fringed with silver leaf paperbark, freshwater mangrove and river pandanus while smaller rainforest communities support slender palms as well as ferns and mosses which hang from the rock walls. The plateau contains larger eucalypts such as the salmon gum and wooly butt as well as hardy golden bark, sand palm, turkey bush and spinifex grass. Late summer and early autumn reveal a brilliant display of heath wildflowers including myriads of ground orchids, lilies and sundries.

Mammals include the antelopine kangaroo, the euro, the agile wallaby, flying foxes, dingos and echidna. The park supports prolific bird life. Especially visible are the noisy red-tailed black cockatoos, parrots, rosellas and the colourful rainbow bee-eater as well as great bowerbirds and bustards. The waterways contain a variety of fish such as barramundi and archer fish while Johnstone's freshwater crocodiles can be seen in winter on the rocks and sandy river banks.

Above and right: *Boat trips enable tourists to appreciate the majesty of the Katherine River and surrounding countryside.*

Dingoes are often seen in Katherine Gorge National Park.

Walking Tracks

There are numerous clearly marked walking tracks beginning from the visitor centre and extending through the park in various directions. Maps and brochures may be obtained from the visitor centre and those who intend to take the camping walks should be prepared with food, water and basic equipment.

Other Activities

Swimming and boat tours down the river are especially popular activities at Katherine Gorge National Park. Fishing, too, is enjoyed either from the bank or from a boat. The various Aboriginal painting galleries are accessible through many of the walking trails.

Climate

From April to October, during the parks dry season, warm sunny days and cool nights are ideal for camping, walking and other activities. From November to March temperatures average 35° C and the high humidity discourages most tourists.

Bookings and Fees

Park entrance and camping fees are payable. Bookings for the dry season can be made through the park ranger.

Information and Bookings

Ranger
Katherine Gorge National Park
C/o Northern Territory Conservation Commission
Gap Road
Alice Springs
Northern Territory 5750
Tel. (089) 72 1886

Ormiston Gorge and Pound

National Park

One of the most striking of the many spectacular gorges which cut through the towering walls of the MacDonnell Ranges is Ormiston Gorge, a deep passage carved in ancient times by the powerful floodwaters of the now sluggish Ormiston Creek. A magnificent permanent waterhole, nearly 100 m long, in the deepest part of the gorge provides a delightful swimming spot on a hot day.

Facilities

Picnic facilities at the entrance to Ormiston Gorge include gas and wood barbecues, showers and toilets. A small camping site is also located here, however it is unsuitable for large caravans. Drinking water should be carried as park water is untreated.

At nearby Glen Helen Lodge motel accommodation, petrol and food are available and the camping site there can accommodate tents as well as all types of caravans.

Bush camping is permitted and if visitors intend to climb Mt Giles, overnight equipment will be necessary. The ranger, located at the information centre near the park entrance, should be consulted before setting out.

Location and Access

The park is situated 132 km west of Alice Springs via Larapinta Drive and the Glen Helen Road. These are sealed

The unusual plumed pigeon is often seen at Ormiston Gorge.

Some excellent bush trails make hiking very popular in the cooler months at Ormiston Gorge and Pound National Park.

but the access road leading into the park is gravel surfaced and may be closed sometimes due to flooding.

Commercial tours from Alice Springs visit the park regularly.

Landscape

Apart from a couple of large waterholes, Ormiston Creek is little more than a string of shallow puddles in the dry season. The creek and another tributary meander over the basin floor from near the entrance of the park towards Ormiston Gorge where the hillsides rise steadily higher and more steeply until they become sheer cliffs glowing red in the sun. The floor of the gorge is strewn with huge rounded boulders which sweep through after summer storms.

Upstream, a rugged natural enclosure known as Ormiston Pound that spans across 10 km is surrounded by high ridges of quartzite which reflect a myriad of colours from the jasper and other silicate compounds forming the high rim. Beyond this, Mt Giles (283 m) is the highest point in the park.

Flora and Fauna

River red gums and bloodwoods dot the dry sandy creek bed while at the far end of the gorge a shaded overhanging wall contains a curtain of ferns. Plant life in this moister microculture includes relict populations from ancient times such as masses of yellow-white daisies that flower in winter and spring. Most of the other plants which survive on the hot and rocky floor of the gorge are sun-loving. Sturt's desert rose, the

The permanent waterhole in Ormiston Creek at the centre of the gorge is a refreshing swimming pool on a hot day.

floral emblem of the Northern Territory, displays delicate pink blooms in the springtime.

The rocky spinifex-covered ridges provide a home for a range of native fauna including wallabies, euros and dingos along with many reptiles. Bird life such as scavenging black kites and the dainty plumed pigeon is abundant. A varied population of fish live in the creek surviving for months when the pools dry out completely by burying themselves in the mud until the next rains.

Walking Tracks

There are no marked walking tracks but visitors may wander throughout the park area. The ranger may be consulted concerning appropriate walks and vantage points.

Climate

Midsummer temperatures range from 23°C to 38°C. Autumn to spring is a more pleasant time to visit the park when daytime temperatures average 19°C. Night time in winter brings very cold temperatures down to as low as 4°C.

Bookings and Fees

There are no entrance fees or camping charges in the park and bookings are not required.

Information

Ranger
Ormiston Gorge and Pound National Park
C/o Northern Territory Conservation Commission
Gap Road
Alice Springs
Northern Territory 5750
Tel. (089) 52 2788

Simpsons Gap
National Park

Only 15 km west of Alice Springs, Simpsons Gap National Park is one of a number of reserves protecting the fauna and wildlife of the red and rugged MacDonnell Ranges. The broken quartzite ridges of Chewings Range, rising 250 m along the eastern wall of Simpsons Gap, form a spectacular chasm above the sandy bed of Roe Creek. Interesting walks provide access to several other gaps and a number of secluded waterholes.

Facilities

At Simpsons Gap, picnic facilities including tables, gas barbecues, parking areas and toilets are available for day use. A visitor centre located at the park entrance offers brochures and displays about the region and the ranger, whose office is situated near the visitor centre, can provide any additional information.

Bush camping is not permitted but a variety of accommodation is available at Alice Springs.

Location and Access

Road access is via Larapinta Drive travelling west from Alice Springs.

Landscape

The park covers 31 000 ha. Running through the middle of the park, Runutjirba Ridge has several gaps including Wallaby Gap, Rocky Gap and Bond Gap, which provide the only access to the northern wilderness. Simpsons Gap itself is a deep chasm cut by a once mighty river running through the Chewings Range. The creek bed is now usually dry but deep permanent waterholes remain in the recesses of the gorges.

Until 1970 most of the park was used as grazing land but since its exclusion the wildlife and fauna is gradually returning to its original state.

Flora and Fauna

The jagged ridges of the park are generally bare of vegetation but the flats between the ranges carry scattered ghost gums, mulga and witchetty bush with a general cover of spinifex in the drier areas. After winter rains there is a brilliant display of flowering annuals.

Rock wallabies, dingos and euros are often seen scurrying about the rocky slopes and plains while bird life includes eagles, kites and numerous other small birds such as the Port Lincoln ringneck. Reptiles are a frequent sight in the park including the harmless children's python and the long-nosed dragon lizard.

Walking Tracks

Various walking trails lead away from each side of the park entrance towards the gaps, waterholes and highpoints. There are a couple of short marked tracks such as the Cassia Hill Walk which provides a convenient lookout point. For further information regarding these the visitor centre provides maps and other details.

Other Activities

Bushwalking, camping, picnicking and sightseeing are the main activities at Simpsons Gap National Park. Swimming in the waterholes is also enjoyed.

Climate

Winter and spring are the best times to visit the park in order to take advantage of the cooler weather and the wildflowers. Summer is extremely hot.

Bookings and Fees

No entrance fees or bookings are required.

Information

Ranger
Simpsons Gap National Park
C/o Northern Territory Conservation Commission
Gap Road
Alice Springs
Northern Territory 5750 **Tel. (089) 52 3131**

Right: *Striking rock formations are characteristic of the broken ridges of the Chewings Range in Simpsons Gap National Park.*

Hiking through the Valley of the Eagles in Trephina Gorge Nature Park.

Trephina Gorge
Nature Park

East of Alice Springs, seasonal tributaries of the Todd River cut two rugged scenic gorges through Trephina Gorge Nature Park. The broad white sand beds of Trephina Creek stand in sharp contrast to the rugged escarpment reaching up to 100 m in height and glowing pink in the sunlight. Towards the west of the park, John Hayes Rock-hole is overhung by broken sandstone walls and huge old eucalypts that offer the visitor shadowy seclusion.

Facilities
Picnic and camping grounds with wood and gas barbecues and toilet facilities are provided at Trephina Gorge. The Ross River Resort has cabin accommodation with private shower and toilet facilities. Bush camping is permitted except around the Ross River area. Water must be carried and the ranger should be consulted first.

Location and Access
Trephina Gorge Nature Park is reached via the Ross Highway, driving east from Alice Springs. Coach tours are available on most days.

Landscape
The park, covering 1700 ha in the east MacDonnell Ranges, features the rugged and secluded Valley of the Eagles, John Hayes Rockhole and Trephina Gorge. It is noted for excellent

The forbidding Valley of the Eagles in Trephina Gorge Nature Park.

scenery which varies from the spinifex hills and bare ridges of the ranges to the shady river beds and waterholes of the gorge floors and the towering broken cliffs which surround them.

Flora and Fauna

In the river bed avenues of whitewoods, ironwoods and ghost gums thrive. The sandstone ledges support hardy desert plants while spinifex and low-growing scrubs cover the higher slopes. The waterholes are thronged year round by seed-eating birds such as the vivid Port Lincoln parrot as well as cockatoos, galahs, kingfishers and magpies. Kangaroos and other native mammals as well as feral cattle may be found throughout the park and reptiles abound in the rocky landscape.

Walking Tracks

Although there are no marked trails, an energetic climb will yield extensive views of the park and the MacDonnell Ranges from numerous vantage points on the hills and ridges.

Other Activities

Picnicking, camping, walking, rock climbing and sightseeing are the chief activities at Trephina Gorge Nature Park.

Climate

Summer is especially hot so the best time to visit the park is in the cooler dry autumn to spring period when the wildflowers are blooming.

Bookings and Fees

Camping fees are payable but bookings are not required.

Information

Ranger
Trephina Gorge Nature Park
C/o Northern Territory Conservation Commission
Gap Road
Alice Springs
Northern Territory 5750
Tel. (089) 50 8211

Uluru (Ayers Rock-Mt Olga) National Park

In the dry heartlands of central Australia the massive monoliths of Ayers Rock and the Olgas dominate the surrounding plains, their colours shifting through blues and violets to browns and fiery reds as the burning sun arcs through the sky. For tens of thousands of years these sandstone mountains have played an important role in the creation mythology of the local Aboriginal inhabitants, and their stories are expressed in many fascinating paintings found in the caves and galleries of the weathered rocks. This is a place of outstanding beauty where the passing of time is measured by the ceaseless trickling of desert sands and the constantly changing moods of the great, brooding Uluru.

Facilities

All tourist facilities are located at the Yulara Tourist Resort, 20 km from Ayers Rock. The camping area occupies 48 ha of undulating sand dune country with accommodation for 3600 people. It has powered caravan sites, tent sites, overnight vans, electric barbecues and shelters, showers, toilets, laundry, swimming pool and kiosk.

The red domes of the Olgas capped by Mt Olga (centre) at 546 m.

Ayers Rock Lodge provides budget accommodation in dormitory-style and family cabins (tel. 089 56 2170). The Four Seasons Yulara Resort (tel. 089 56 2100) and Sheraton Ayers Rock Hotel (tel. 089 56 2200) provide international standard accommodation.

Other facilities at Yulara include a visitor centre, lookouts, bank, restaurants, bars, sporting facilities, a medical centre, supermarket, bicycle hire, car rental and a service station.

The ranger station, toilets and picnic area are close to Ayers Rock. Camping is not permitted within the park boundaries.

Location and Access

The park is 465 km south-west of Alice Springs via the sealed road from Erldunda. There are regular flights, express coaches and coach tours from Alice Springs. Scenic flights and other tours can be arranged in Yulara. The road to Ayers Rock is sealed; the road to the Olgas though unsealed is usually suitable for conventional vehicles.

Landscape

The park occupies 126 132 ha on the edge of the Gibson and Great Victorian Deserts. The surrounding landscape is one of mainly flat and undulating sandplains, dunes and mulga woodland, approximately 500 m above sea level. Ayers Rock, the world's greatest monolith, rises 348 m above ground level, and has a circumference of more than 9 km. Worn down by the elements, its composite, dull-red sandstone structure conserves the water from the occasional downpour in pools.

Deep grooves along the sides of Ayers Rock are a result of rainfall erosion.

Above: *A bustard at Uluru National Park.*

Left: *Ayers Rock, the world's greatest monolith.*

Like the Olgas it is the weathered tip of an underground, steeply tilted rock strata formed when the area was submerged some 500 million years ago. Over the millennia the softer surrounding rocks have been eroded away to leave these great monoliths towering over the plains.

The Olgas consist of more than 30 red, dome-like rocks covering an area with a circumference of about 30 km, and separated by deep canyons. The largest is Mt Olga, rising 546 m above ground level.

Flora and Fauna

Despite the low rainfall and barren appearance of much of the area, there is a surprising variety of flora and fauna. On top of and around the rocks are euros, rabbits, dingos, foxes and bats, domestic mice, hopping mice and marsupial mice, the endearing red-eared antechinus, echidna, destructive feral cats and rats. Amongst the mulga and on the plains are great red kangaroos and many camels.

Reptiles are common and include king and western brown snakes, the children's python, the enormous perentie lizard, skinks, sand goannas, geckos and the fascinating thorny devil which feeds on the many ants.

High on the rocky ledges and caves wedge-tailed eagles and nankeen kestrels make their nests. Among the mulgas are honeyeaters, masked woodswallows, zebra finches, diamond doves and crested pigeons. Bustards, emus and flocks of budgerigars are often seen on the plains, while black crows scavenge the countryside.

Around the rocks are fig trees, native plums, hop bushes, bloodwood gums, wild oats and swaying kangaroo grass, with ferns and trigger plants hidden in cracks and crevices. Spinifex dominates the plains with clumps of mulga protecting parakeelyas and white daisies that bloom prolifically after the rains. Magnificent stands of desert oaks appear among the spinifex, while on the sandhill crests are brilliant grevilleas and many small pea plants. Red river gums line some of the watercourses of the Olgas, together with Sturt's desert rose and the unique early nancy which produces a pink carpet of flowers when conditions are right.

Walking Tracks

Marked walking tracks lead around Ayers Rock and the Olgas. Most people follow the track to the top of Ayers Rock, which takes about two hours return and is quite strenuous. Tracks around the Olgas lead through gorges, to Kata Tjuṯa Lookout and into the Valley of the Winds. There are many opportunities for bushwalking around the Olgas and into the sandy desert areas, but the ranger should be consulted before leaving, and water must be carried.

Rangers conduct regular tours discussing the park's wildlife, geology, history and Aboriginal mythology.

Other Activities

The changing colours of the rocks and desert landscapes offer plenty for the photographer, and those interested in Aboriginal mythology will find the area one of the richest in Australia.

Climate

With an average annual rainfall of around 200 mm, the weather is usually fine. Summer is extremely hot, with temperatures reaching 44°C. The best time to visit is from May to October when there are warm days and cool nights, although the night temperatures can drop below freezing point in July.

Bookings and Fees

Bookings for all Yulara accommodation can be made through any office of the Northern Territory Government Tourist Bureau. Small camping fees and a park entry fee are payable.

Information

Visitor's Centre
Uluṟu National Park
P.O. Box 119
Yulara
Northern Territory 5751
Tel. (089) 56 2988, 56 2980

Western Australia

Western Australia is a region of diversity. In the north-west are landscapes of iron-red ranges, immense gorges and river beds edged with shady gums. In the south are the sandplains, the forests of towering karri, the small jagged ranges, and the scenic south coast of bold granite headlands and lonely white beaches.

In the far north are the National Parks of the Kimberley region—mountain ranges formed millions of years ago. The largest of these parks is Geikie Gorge where multi-coloured cliffs are reflected in the placid waters of the Fitzroy River.

The parks of the north-west include arid ranges, huge gorges and lush river scenery. Within this area are the Hamersley Range, Millstream-Chichester and the coastal Cape Range National Parks.

North of Perth, the wild beauty of landscapes is unsurpassed at Kalbarri National Park and unusual rock formations stud the flat, sandy open space at Nambung National Park. Yanchep National Park was extensively developed as a recreational park before it came under National Parks' management.

National Parks in the Perth region include John Forrest and further south, Leeuwin National Park.

The south coast—where the oldest landforms confront the stormiest ocean—is an area of bold contrasts. Here, windswept sandplains give a foretaste of the Nullarbor; jagged ranges overlook the boiling white foam of the Southern Ocean and towering forests crowd sheltered inlets. The National Parks of this region include Walpole-Nornalup, Stirling Range, Fitzgerald River, Cape Le Grand and Cape Arid.

Spectacular wildflower displays are perhaps the best-known feature of Western Australia's parks. The state has more than 3000 species which are best seen from August to October.

Cape Arid
National Park

Cape Arid National Park contains magnificent coastal scenery, with fingers of granite sheltering shallow, sandy bays where swimming and fishing are popular. Coastal dunes and swampy flats run inland to an area of sandplain and the northern wooded Russell Range. The reserve is of great botanical interest as it contains the flora examined by botanist Robert Brown who accompanied Matthew Flinders on his voyage of exploration in 1801.

Facilities
The Thomas River campsite has car access, toilets, tables and water available for bathing and washing. Mount Ragged and Thomas Fishery campsites are all four-wheel-drive access and have toilets. Jorndee Creek, Seal Creek and Poison Creek campsites have similar facilities but are accessible by two-wheel-drive vehicles on gravel roads. Visitors must bring their own drinking water and gas barbecues.

Bush camping is permitted but campers should inform the ranger of their intentions as the park is largely waterless. The ranger is resident near the mouth of the Thomas River.

Location and Access
Cape Arid National Park is located 125 km east of Esperance. The park is situated on the south coast of Western Australia between Israelite Bay and Cape Le Grand. Car access is from Fisheries Road. From the sealed section of this road, 19 km of all-weather gravel road leads to the main camping area at Thomas River. The majority of tracks within the park are rough four-wheel-drive and access is subject to weather conditions.

Sand dunes along the coast in Cape Arid National Park.

Landscape
The 279 415 ha park has impressive coastal scenery. Beaches of Yokinup Bay make a sandy sweep of 20 km from the mouth of the Thomas River to the granite headland capped by Mount Arid (356 m). The peak of this mount gives fine views of the shoreline and dozens of islets of the eastern Recherche Archipelago. Coastal dunes and swampy clay flats border the shoreline and inland the park consists mostly of sandplains. In the north is the wooded Russell Range, where razorback Mount Ragged rises to 585 m. Several ruins and small mounts provide superb outlooks.

Flora and Fauna
Young sand dune systems support coastal heaths with patches of yate, banksia and paperbark. Mallee stands, sometimes dense, occur in the northern areas. Cape Arid has many wildflower species, some being magnificent in flower and found only on these granite headlands.

Diverse habitats within the park ensure a wide range of fauna. Some of the more unusual species easily sighted include honey possums, brush-tailed wallabies, Cape Barren geese, sea eagles and bustards. Several species of reptiles are also found and there are some magnificent ant beds in the Belinup Hill area.

Walking Tracks
The coastal track offers superb views of the many islets. The walking is hard with some steep climbs. There is a well-marked native trail of 1 km near the camping ground of Belinup Hill.

Other Activities
Swimming can be enjoyed in the Thomas River. Most beaches are suitable for swimming, but caution is necessary as rips can develop quickly. Rock climbing is allowed with permission from the ranger. Picnicking and observing the wildflowers are also enjoyable activities within the park.

Climate
The best times to visit are from October to May, for both the wildflowers and pleasant weather.

Bookings and Fees
There are no fees payable and bookings are not required.

Information
Ranger-in-charge
Cape Arid National Park
P.O. Box 185
Esperance
Western Australia 6450
Tel. (090) 75 0055

Calm District Office
P.O. Box 234
Esperance
Western Australia 6450
Tel. (090) 71 3733

Granite headlands tower over long stretches of sandy beaches in Cape le Grand National Park.

Cape Le Grand
National Park

Less than one hour from Esperance, on the southern coast of Western Australia, Cape Le Grand National Park combines the dramatic coastal scenery of granite hills and headlands with peaceful pure white sandy bays.

Facilities
Caravan and car camping areas are located at Le Grand Beach and Lucky Bay, where there are fireplaces, toilets, hot showers and drinking water. Bush camping is not generally permitted throughout the park, but the ranger should be consulted about beach camping sites.

There is a ranger's residence and information bay near the park entrance on Cape Le Grand Road.

Picnic areas are located at Le Grand Beach, Hellfire Bay, Thistle Cove, Lucky Bay and Rossiter Bay. There are boat launching ramps near the campsites.

Accommodation is available in Esperance and at the caravan park at nearby Duke of Orleans Bay.

Location and Access
The park is 47 km east of Esperance, access is via Merivale Road, turn off into Cape Le Grand Road. Roads are being sealed and there is bitumen along 8 km of the Le Grand Road to Lucky Bay Road. Lucky Bay Road is currently being sealed. Day tours operate from Esperance.

Landscape
The park covers an area of 31 390 ha. The coastline consists of wide sandy beaches between rocky headlands backed by massive granite outcrops in the south-western corner forming a chain of peaks rising to 352 m at Mount Le Grand. Further inland are undulating sandplains covered with rich coastal heaths, swamp and freshwater pools. Several large caves run under the granite peaks, the most spectacular being a large hole under the summit of Frenchman Peak.

Flora and Fauna
Grey kangaroos, rock wallabies, pigmy possums, honey possums and bandicoots inhabit the park.

More than 100 species of birds have been recorded in the park, among these are many sea birds including penguins, petrels, shearwaters, Pacific gulls and albatrosses. Among the coastal heaths are honeyeaters, the western emu wren and the red-eared finch which is found only in south-western Australia.

Left: *The white sandy shores and crystal waters of Lucky Bay provide an idyllic camping site.*

Below: *Ruins of the old Telegraphic Station at Eucla in south-east Western Australia.*

The sandplains support dense thickets of banksias, with paperbarks in the swampy areas, while on the granite hills are several species of mallee. There are many acacias, blackboys, kangaroo paws and native honeysuckles, with beautiful wildflower displays in spring.

Walking Tracks

The spectacular coastal walking trail provides excellent walks varying from less than one hour to a whole day, passing the many beaches, headlands and granite peaks. The track is well marked, although water should be carried in the summer months. Other tracks lead along the beaches and hillsides.

Other Activities

The many beaches offer good swimming and surfing possibilities, but beware of rips. Boating is popular, and there are many good fishing spots.

Climate

The park has a very mild climate with cool, wet winters and warm, dry summers. Visit all year round.

Bookings and Fees

Bookings are not accepted for the camping area. Camping fees are payable.

Information

The Ranger
Cape Le Grand National Park
P.O. Box 706
Esperance
Western Australia 6450
Tel. (090) 75 9022

Cape Range
National Park

Cape Range National Park is an area of tranquil sandy beaches, rugged ranges and magnificent canyons. Some rain falls here in winter and cyclonic storms may flood the gullies in summer, but for most of the year the land is parched, the deep blue water of Yardie Creek, which lies imprisoned between vertical cliffs, being the only permanent water and, as such, giving an impression of a semi-tropical fjord. Camping is the main attraction of this scenic park. The camps are strung along a rough vehicle track that runs the full length of the ocean boundary. A new road is being constructed to Yardie Creek and connecting roads lead to the campsites.

Facilities

Numerous campsites are located along the coast and camping is limited to twenty-eight nights in any one spot. Camping facilities are rudimentary. Visitors are advised to bring chemical toilets and gas cookers. No fires are allowed in the park and barbecues are not provided. Fresh water is available at Mangrove bore near the northern park entrance and at the Milyering Information Centre.

From May to October a ranger lives near the northern entrance to the park on the west coast and there is another ranger stationed near Yardie Creek. The resident ranger is based at Exmouth.

The sandy beaches of Cape Range National Park stretch for 50 km.

Location and Access

Cape Range National Park is located 400 km north of Carnarvon. Car access is off the Learmonth–Exmouth road or 40 km past Exmouth. Within the park some sections of the road are rough but, if taken with care, access is generally available. Immediately after rain access is limited but signs at Exmouth and the park boundary give warning at these times.

Landscape

The 50 000 ha park is situated on the western side of the North West Cape. The Cape Range, backbone of the Cape, is a huge upward fold of limestone originally formed below the sea. Erosion has gouged vast canyons in the ranges and deposits of eroded material have built up along the coastal sides of the range. The park's coastline, made up of stretches of sandy beaches, reaches for 50 km.

Flora and Fauna

There are a number of plants unique to the Cape and botanists will be surprised to find plants normally associated with the south-west of the state growing here. Examples are hibbertia and banksia, the former often growing amongst spinifex tussocks. Mangroves, acacia, cassia, spinifex species on the gorge slopes, minilya lilies, grevillea, eucalyptus and verticordia are also found in the park. Kangaroos are common and very tame. Rock wallabies, euros, 80 species of reptiles and over 100 species of birds also inhabit the area.

Walking Tracks

Cross-country walking is not advised, but safe trails are marked along a short section of the range crest, giving good views of the eastern canyons and the gulf. These are reached by two roads turning inland between Learmouth and Exmouth. The area is rugged and waterless and temperatures in the gorge can be much higher than elsewhere. Walking in summer can be dangerous and walkers are always advised to let somebody know of their plans. Brochures are available from the rangers or the Exmouth Tourist Bureau.

Other Activities

Swimming, fishing, scenic sightseeing and birdwatching can all be enjoyed within the park. The park was the site of Australia's first oil strike—the capped well head can be seen on Charles Knife Road.

Climate

May to October is the best time to visit. The winter months are mild with a little rain early in the season. November to April is cyclone season with very hot temperatures. If visiting the park during cyclone season, it is essential to carry a radio to monitor weather reports. The park is at its most attractive a month or two after a cyclonic rainstorm.

Bookings and Fees

Camping and park entrance fees are payable but bookings are not required.

Information

Resident Ranger
Cape Range National Park
P.O. Box 201
Exmouth
Western Australia 6707
Tel. (099) 49 1428
Fax. (099) 49 1580

District Manager
Gascoyne Region
Department of Conservation and Land Management
P.O. Box 201
Exmouth
Western Australia 6707
Tel. (099) 49 1676
Fax. (099) 49 1580

Gum flowers add to the wildflower display in Fitzgerald River National Park.

Fitzgerald River
National Park

West of Esperance, where the Fitzgerald and Hamersley Rivers cut through the rugged Mount Barren Ranges to the ocean, is one of the loveliest stretches of the south-west coast of Australia. Here the Fitzgerald River National Park preserves an area of craggy cliffs and tall rock spires pounded by the Southern Ocean, sheltered bays, long sandy beaches stretching as far as the eye can see, and vast sandplains which nourish and protect many unique and unusual wildflowers.

Facilities
Camping is possible at Mylies Beach, 15 km west of Hopetoun, where there are toilets. Bush camping is generally permitted throughout the park, but water should be carried as the rivers are salty, and the ranger consulted about suitable bush camping sites.

Left: *Binding grasses protect the beach in Fitzgerald River National Park.*

There are rangers' residences at Barrens Beach between Hopetoun and Mylies Beach, near Gordon Inlet in the south-west of the park, and at Jacup.

Accommodation is available in the towns of Bremer Bay, Hopetoun, Jerramungup and Ravensthorpe.

Location and Access
The park is 240 km west of Esperance and 225 km east of Albany. Access from the east is via Ravensthorpe to Hopetoun, turn off west along Hamersley Drive, 1 km before Hopetoun. From the west turn off the South Coast Highway about 30 km before Jerramungup into Devils Creek Road or into Quiss Road, 20 km east of Jerramungup. Tracks from Bremer Bay into the park are very rough.

Roads within the park are all unsealed, and apart from the above access roads are suitable for four-wheel-drive vehicles only.

Landscape
The park covers an area of 242 739 ha consisting mainly of undulating sandplains with wide, flat valleys and narrow gorges carved out by such rivers as the Hamersley, Fitzgerald, Phillips and Gairdner. The rugged Barren Mountains thrust up isolated peaks in places, rising to 529 m. East of the Fitzgerald Inlet the mountains drop off into the ocean in a series of massive cliffs interspersed with small sandy inlets. West of the Fitzgerald Inlet rocky headlands thrust out from long stretches of sandy beach.

Flora and Fauna
Western grey kangaroos are frequently encountered in the park, other mammals include brush wallabies, tammars, honey and pigmy possums, bush rats and short-nosed bandicoots, while sea lions can sometimes be seen lounging on the rocks.

Bird life is prolific and colourful. There are large emus and tiny emu-wrens, parrots, rainbow birds, golden whistlers and blue wrens; brown hawks are common, and the wedge-tailed and white-breasted sea eagles can be seen riding the thermals. Albatrosses, shearwaters and petrels are common along the coast.

There are many goannas, geckos, frogs and snakes, but they usually keep well out of sight.

The park has an enormous variety of plants with beautiful wildflower displays in late winter and spring. There are wisp-like weeping gums, royal hakeas, many low shrubs like the grevilleas, acacias, and isopogons. Taller shrubs include mallees, banksias and the Christmas tree. In the valleys are stands of round-leaved mort and pincushion hakeas, whereas in the swampy woodlands are swamp yates and paperbarks.

Walking Tracks
There are few established walking tracks, the main one being from Point Ann to East Mount Barren with superb coastal views. This is intersected with four-wheel-drive tracks and can take a number of days to complete. There are bushwalks to hilltop viewpoints and scenic wildflower walks. Water must be carried.

Other Activities
Beach and rock fishing are popular. There are many places to swim and surf, although rips are common; boating is considered dangerous due to rapidly changing conditions. The mountains and rock faces offer good climbing possibilities.

Climate
The park has warm summers and cool winters, with most rain falling in the winter. Visit all year round.

Bookings and Fees
Bookings are not accepted for the camping areas. Camping fees are payable.

Information
Ranger-in-charge
Fitzgerald River National Park
P.O. Box 33
Jerramungup
Western Australia 6337
Tel. (098) 35 5043

Impressive multi-coloured limestone cliffs at Geikie Gorge are scoured white along the lower levels by the torrential wet season floodwaters of the Fitzroy River.

Geikie Gorge
National Park

In the far north of Western Australia, where the Kimberleys rise from the Great Sandy Desert, is a great barrier reef system that grew and flourished below the warm Devonian oceans some 350 million years ago. Left high and dry by the receding waters, this remarkable geological structure survives in the shape of the massive rock walls of the Napier, Oscar and Geikie Ranges. Over the years great rivers have carved steep-sided, multi-coloured gorges through the limestone rock. The best known is the Geikie Gorge, now a National Park, and home to an amazing variety of wildlife including living specimens of ancient ocean fish.

Facilities
An extensive caravan and car camping area is located at the gorge entrance with fireplaces, picnic tables, toilets and showers. The camping area is, however, flooded during the summer. Accommodation is available at Fitzroy Crossing. Bush camping is not allowed in the park.

Location and Access
The park is some 2700 km north-east of Perth, 1500 km south-west of Darwin and 16 km north of Fitzroy Crossing. Turn off the Great Northern Highway on the unsealed road to the park. This road is impassable after heavy rainfall.

Landscape
The park covers 3136 ha and extends 8 km along the Fitzroy River. The Geikie Gorge is about 400 m wide with steep, colourful limestone sides rising some 60 m above the water. The river has carved fantastic shapes from the rock faces, scoured white by the rushing torrent in the wet season, stained red by iron oxides, and weathered grey at the upper edges. In the dry season the river falls by some 16 m leaving a series of beautiful, tranquil waterholes between sandbanks.

Flora and Fauna
Among the mammals in the park are agile wallabies, euros, rock wallabies and large colonies of flying foxes along the river bank. More than 100 bird species have been identified. Around the river are darters, cormorants, parrots, cockatoos, pigeons, herons, sea eagles, spoonbills, kingfishers, the ibis and greater bowerbird. In the caves are colonies of fairy martins with their distinctive mud nests.

Estuarine crocodiles laze on the rocks, while archer fish circle just below the surface hunting insects which they shoot down with a water jet. There are barramundi and, exiled hundreds of kilometres from the ocean, sawfish and stingrays.

The dry hillsides with open savannah vegetation contrast with the dense riverine forests on the gorge. There are many river gums and cajeputs here with native figs, freshwater mangroves, Leichhardt trees, screw pines and passionfruit vines scrambling over them. The tall tropical reed, *Phragmites karka,* is common.

Walking Tracks
A walking track leads along the west bank of the river. The rest of the park is difficult terrain for walkers.

Other Activities
The best way to see the gorge is by boat, and the ranger conducts boat trips through the gorge twice daily from April to November. The use of private boats is strictly controlled,

and they may only follow the tour boat in restricted numbers. Swimming is only permitted on the west bank, south of the camping area.

Climate

Summers are hot and wet, winters warm and dry. Visit between April and November.

Bookings and Fees

No bookings are taken, but camping fees are payable. Private boats must pay a launching fee.

Information

The Ranger
Geikie Gorge National Park
C/o P.O. Fitzroy Crossing
Western Australia 6765
Tel. (091) 91 5030

Right: *The darter or snake bird, a strong underwater swimmer who lives on fish, frogs and yabbies in Geikie Gorge.*

Below: *Tunnel Creek runs through an eerie underground passage for 750 m through the Oscar Range in the West Australian Kimberleys.*

Hamersley Range
National Park

In the heart of Australia's great western mountain chain, Hamersley Range National Park encompasses an area so ancient that its rocky ramparts and deep gorges reveal rocks formed 2000 million years ago, when the earth's crust first settled and solidified. In the glaring sunlight these rocks glow in hues so vivid that they appear to be reflecting the very primal forces that created them. This is a huge, hot, dry land, where summer cyclones suddenly flood the sparsely vegetated plateau, cutting great chasms, raw ridges and rugged bluffs of incredible colours and stunning grandeur.

Facilities
There are four caravan and car camping areas in the park with free gas barbecues, picnic tables and bush toilets. The only permanent water is in the gorges, and visitors must not pollute the pools with soap. Drinking water should be brought in. Accommodation is also available at Wittenoom.

Ranger headquarters are located on the Juna Downs –Yampire Road within the park.

Location and Access
The park is 1500 km north of Perth via the Great Northern Highway, turn off on the unsealed road for Wittenoom before Roy Hill. Port Hedland is 295 km north along an unsealed road. Roads within the park are good but unsealed.

Landscape
The park covers 617 606 ha of a high, hilly plateau, rising to 1235 m at Mount Bruce in the central area. Spectacular gorges, some very deep and narrow, have been etched into the plateau by creeks feeding into the Fortesque River to the north of the park. The rock formations lie in clearly defined layers, glowing with deep reds, browns, mauves and blues, and contrasting sharply with the dry, sparse, yellow spinifex of the plateau. Many creeks cross the plateau before plunging in tumbling cascades into the cool gorges where lush vegetation grows in the shadowy depths.

Flora and Fauna
Euros, the very shy brush-tailed rock wallaby, echidnas, the pebble-mound mouse, dingos and many species of bats inhabit the park. Pythons, goannas, geckos and legless lizards live among the rocks. The park has a wide variety of bird life particularly around the river pools, including reed-warblers, finches, kingfishers, rock pigeons and wrens.

Spinifex grass covers the rolling hills, with white-barked snappy gums on the hillsides, mulga trees are widespread, and giant river gums and cajeputs line the gorges. The rock pools are fringed with rushes with ferns growing on the moist, shaded banks of the deep gorges. Many wildflowers bloom in the winter when flowering cassias, acacias, northern bluebells and mulla mullas abound.

Left: *In the dry season boat tours down the tranquil Fitzroy River enable visitors to appreciate the splendour of Geikie Gorge.*

A goanna enjoys the sunshine in Hamersley Range National Park.

Walking Tracks
The park has many walking tracks leading into magnificent gorges; Kalamina Gorge and Dales Gorge, Weano and Hamersley Gorges being particularly beautiful and easily accessible. A path leads to Mount Bruce; there are cliff-top walks and short walks through rocky spinifex country. Many long expeditions can be made through the gorges, although they may involve some swimming, and the ranger-in-charge must be consulted.

Many huge termite mounds are to be seen above the gorges in Hamersley Range.

The cool waters of Circular Pool, a permanent waterhole which traps the wet season waters of the Fortesque River at Dales Gorge in the Hamersley Range National Park.

Other Activities
There are excellent photographic opportunities in the park and many swimming spots in the gorges.

Climate
November to April is very hot with most of the 250–300 mm of rain falling at this time. Winters are warm with cold nights. Visit from May to October.

Bookings and Fees
Camping fees are charged but bookings are not accepted.

Information
Ranger-in-charge
Hamersley Range National Park
P.O. Box 20
Wittenoom
Western Australia 6752
Tel. (091) 89 8157

John Forrest
National Park

The Perth skyline can be seen from this charming, heavily wooded park on the crest of the Darling Range escarpment. First called Greenmount, the area was known from 1900 simply as 'the National Park'—Western Australia's first. John Forrest is heavily used in the warm spring weather and prone to fires in the summer.

Facilities
There are limited camping facilities and only backpack camping is allowed. A permit is needed for bush camping. Centrally located tea rooms and a taverne/restaurant provide excellent facilities for day visitors. Jane Brook has been dammed to create a natural swimming area and there are

toilets, shelter and change rooms alongside the pool. Nearby is a children's playground, a picnic area and barbecue facilities. Another popular picnic area is situated near Rocky Pool on the western side of the park.

Location and Access

John Forrest National Park is located 25 km east of Perth. The park adjoins the Great Eastern Highway between Greenmount and Glen Forrest and there are three points of access. There is a vehicle access fee. Foot access is from Swan View or Hovea.

Landscape

Natural features in this 1580 ha park suffered heavily under early policies aimed at taming its streams, particularly Jane Brook, and turning the surrounds into English-style pleasure grounds. However, the jarrah forest is still largely in its natural state. The park is on the crest of the Darling Range escarpment and is covered in typical Darling Range forest growing on rocky slopes. Open woodlands spread along the inland side of the scarp and granite outcrops are dotted throughout the landscape. Peaceful streams, as well as a disused railway line, cross the park.

Wallabies are often sighted in John Forrest National Park.

Flora and Fauna

Jarrah, bull banksias and red gum are common and wandoo may be found in many parts of the park. Unfortunately some areas are infected by dieback disease which has killed some of the jarrah, bull banksias and undergrowth species. Wildflowers, especially in spring, are profuse and include blue leschenaultias, hoveas, red and green kangaroo paws, pink myrtles and many orchids. There are beds of cultivated wildflowers with name plates to enable easy identification. Blackboys and zamias add variety to the scene.

The trees and wildflowers attract prolific bird life. Kangaroos and echidnas, short-nosed bandicoots and wallabies are often sighted and there is a large range of reptiles.

Walking Tracks

There are extensive walking tracks throughout the park. Some lead to attractive features such as Hovea Falls, National Park Falls, Mahogany Creek Dam and Glen Brook Dam. One interesting route is along the disused railway line. The tracks are generally short but hilly.

Other Activities

The park is a popular venue for all the family. Picnicking, sightseeing from lookouts and swimming may all be enjoyed and there are short, guided, wildflower tours in spring.

Climate

Visit from July to October for most of the wildflowers, otherwise any season.

Bookings and Fees

There is a vehicle access fee.

Information

The Supervisor
John Forrest National Park
C/o P.O. Glen Forrest
Western Australia 6071
Tel. (09) 298 8244

Kalbarri
National Park

On the west coast of Australia, north of Geraldton, the Murchison River has cut a deep, serpentine channel through the elevated sandplains. After the summer cyclones the river becomes a raging torrent, forcing its way along great gorges through multi-coloured rocks to pour into the Indian Ocean amidst cliffs of gold, white and brown. Kalbarri National Park protects the final and most spectacular part of this great river journey—a gorge 80 km long and 170 m deep.

Facilities

Formal camping is not allowed in the park, although there are six caravan and camping parks with all facilities in the township of Kalbarri. Motel and hotel accommodation, cottages and units are also available here.

Hikers are allowed to camp overnight in the park, but written permission is required from the ranger, and there must be at least five people in the party. Wood fires are not permitted.

Picnic areas and toilets but no water are located along the gorge. There is a free gas barbecue at the Z-Bend picnic area.

Shaped by windblown sand, Nature's Window frames a spectacular view of Murchison Gorge.

Car parking areas are provided at the viewing spots along the coastal cliffs, and there is an information board at the entrance to the park near Kalbarri.

Location and Access

The park is some 590 km north of Perth on sealed roads. Turn off the North West Coastal Highway towards Kalbarri approximately 155 km north of Geraldton. Access roads within the park are mainly unsealed. Regular coach tours to the park's scenic areas are conducted from Kalbarri.

Landscape

The park covers 186 096 ha, its main feature being the winding Murchison Gorge which cuts through soft sandstone some 170 m thick, revealing all the rich colours of the Australian heartlands, in a succession of tortuous loops and bends. Rolling sandplains comprise the majority of the park, supporting small trees and a great variety of wildflowers. They end abruptly in colourful sandstone cliffs and headlands falling in giant steps to the thunderous ocean below.

Flora and Fauna

Red and western grey kangaroos, euros and rock wallabies inhabit the park. Emus are often seen on the sandplains, while more than 170 other bird species have been recorded. They include many waterbirds such as waders and ducks. Kestrels nest on projecting ledges in the gorges. Honeyeaters, finches, parrots and rosellas are common.

The sandplains support more than 300 species of flowering plants, with massed stands of golden verticordia, pincushion plants, banksias, grevilleas, melaleucas, claw flowers, kangaroo paws and blackboys. Acacias provide beautiful springtime displays, native cypresses and eucalypts are common, and great river gums line the gorge.

Walking Tracks

There are many short walks down to the gorge from lookouts, revealing spectacular scenery, rock pools and sandbanks. Numerous paths lead from the road to cliffs with such classical features as sea arches, sea stacks and coastal gorges. A nature trail leads to Mushroom Rock.

Overnight hikers in the river gorge must inform the ranger before departing, have a minimum of five people in the party, and report back to the ranger on completion of the journey.

Other Activities

Fishing and swimming in the river estuary are popular activities. It is possible to canoe in the river gorge, but the ranger should be informed before departure.

Climate

The park has mild winters and hot summers. Visit all year round.

Bookings and Fees

A park entrance fee is payable.

Information
The Ranger
Kalbarri National Park
P.O. Box 37 Kalbarri
Western Australia 6532
Tel. (099) 37 1140

Left: *High cliffs of sandstone weathered by wave action and winds create many unusual rock formations in Kalbarri National Park.*

Right: *North of Geraldton at Kalbarri National Park the West Australian coastline presents a deeply dissected, forbidding vista.*

Leeuwin
National Park

A spectacularly rugged coast between the two capes that jut out from the south-west corner of the continent; a hinterland ranging from low heaths to huge karri forests; and extensive cave systems exquisitely decorated with intricate formations are found in Leeuwin National Park. Some of the caves contain fossil deposits, including marsupials no longer found here—the Tasmanian tiger and the Tasmanian devil—and extremely old human bone fragments.

Facilities
There are no designated campsites but bush camping is permitted subject to approval from the ranger. Picnic facilities with toilets and barbecues are scattered throughout the park.

Outside the park there are hotel/motel and camping facilities at Yallingup, Margaret River and Augusta; and there are camping facilities at Cowaramup Bay and Hamelin Bay.

Rangers are on duty in the Yallingup, Cowaramup and Leeuwin areas.

Location and Access
The park is located 270 km south of Perth. Car access is via the Bussel Highway with many entries along Caves Road.

Landscape

Leeuwin National Park, comprising 15 493 ha, consists of a series of reserves on a 80 km ridge of granite gneiss and limestone. Nowhere is the Indian Ocean coastline more varied than here, at its beginnings. Wide beaches and dunefields are interspersed with rocky headlands and sections of high, windswept cliffs. Nearshore islets and reefs are strewn along much of the coastline. Inland, water permeating the limestone has hollowed out caves such as the Yallingup, Lake Mammoth and Jewel. The Cape Leeuwin lighthouse and waterwheel are included within the park's area.

Flora and Fauna

Within such a varied environment there occurs a great variety of flora and fauna. The vegetation ranges form wind-stunted heaths and mallee scrubs to eucalypt forests, including tall stands of jarrah in the south. Paperbark woodlands occur in wetter areas, or in the thickets on sheltered parts of the coast. Over 200 species of flowering plants have been recorded and orchids are numerous. Two particularly rare plants are found here: the *Kennedia macrophylla*, a large-leaved creeper with red flowers, and *Eucalyptus calcicola*, a small, mallee-like eucalypt.

Water rats are only one of the many mammals found in Leeuwin National Park.

Wildlife is also varied. Mammals found include the western quoll, western water rat, and ring-tailed possum. Amongst the bird life is the extremely rare rufous bristle bird, the nearest of its kind are found in South Australia. The red-tailed tropic bird nests on Sugarloaf Rock and the osprey, rare in many parts of the world, patrols the coast and nests on the rocky islets or tall trees just inland. The seascape provides an abundance of sea birds and sometimes a whale can be seen breaching the surface.

Walking Tracks

There are several walking trails which provide short walks and fine scenic panoramas. The more adventurous can walk the 100 km trail along the top of the granite cliff ridge. There are marked trails running between parks.

Other Activities

There are a number of safe swimming beaches along the coast, including excellent surf beaches. Fishing, boating and visiting the caves are also popular activities.

Climate

Spring to autumn are the best months to visit.

Bookings and Fees

No fees are payable.

Information

Leeuwin Area
Ranger
Leeuwin National Park
P.O. Augusta
Western Australia 6290
Tel. (097) 58 1756

Yallingup Area
Ranger
Leeuwin National Park
R.S.M. 435
Busselton
Western Australia 6280
Tel. (097) 55 2144

Cowaramup Area
Ranger
Leeuwin National Park
P.O. Cowaramup
Western Australia 6284
Tel. (097) 55 5324

Millstream-Chichester National Park

A staircase series of deep spring-fed pools along the Fortesque River lined by river red gums, tropical paperbarks and tall fan palms provide an enticing oasis in the midst of the hot, dry Pilbara. The boundaries of Millstream National Park have recently been enlarged to join the Chichester Range, a stony, sparsely wooded region relieved by ephemeral creeks and cool gorges. The two now form an area of 200 000 ha known as the Millstream-Chichester National Park.

At Millstream-Chichester National Park permanent spring-fed pools surrounded by lush vegetation form an oasis in the often dry Fortesque River valley.

Facilities

At Millstream there are three picnic sites with fireplaces and tables. There is a camping area with toilets, water and caravan sites near the Deep Reach Pool, and Millstream Tavern, providing refreshments, is located near the ranger's residence and Chinderwarriner Pool. Another camping site is located at Snake Creek in the Chichester Range with water and toilet facilities.

Bush camping is permitted at the ranger's discretion.

Location and Access

The park is situated 250 km south-west of Port Hedland. Car access is off the North-West Coastal Highway via the Wittenoom Road which traverses the park.

Landscape

The original valley of the Fortesque River has filled with porous sediments, forming an aquifer which feeds the pools at Millstream and supplies them with permanent water. The Fortesque River itself often dries up. To the north-east, beyond the Fortesque floodplain, the landscape is characteristic of the arid Pilbara. Harsh, rocky country with a ground cover of grasses is eroded by numerous creeks with some gorges and permanent waterholes.

The Chinderwarriner Pool and the Millstream area in general is of particular significance to the Aboriginal people.

Flora and Fauna

Because of the permanent water supply, Millstream supports a variety of plants and wildlife. Apart from river gums, cajeput and Millstream palms, the remains of early pastoralist activity can be seen in surviving date palms, cotton palms, oleanders, bamboo and waterlilies. Bullrushes are found around the pools. The Chichester Range has sparse eucalypts and a general cover of hummock grasses, with dense riverine vegetation in the gorges.

Many black flying foxes, the only large colonies in inland Pilbara, as well as euros and red kangaroos, are found in the region. Waterfowl, warblers and tree creepers are prolific while many species of fish including the spangled perch are found in the pools and waterways.

Waterlilies thrive in the cool damp gorges of Alligator Creek in Millstream-Chichester National Park.

Walking Tracks

There are a number of easy, well-marked trails around the Chinderwarriner Pool area and from Snake Creek in the Chichester Range, a 20-minute walk to Python Pool offers a good sample of the area's gorge scenery.

Other Activities

Swimming, camping, picnicking, bushwalking and sightseeing are the chief activities in the park. Swimming is discouraged in Chinderwarriner Pool but the foundations of an old bath house built over one of the channels remain.

Climate

The area is very hot in the summer months but from April to October the climate is pleasant and warm.

Bookings and Fees

There is no park entrance fee but a camping charge is payable. Advance bookings are not required.

Information

The Ranger
P.O. Box 119
Karratha
Western Australia 6714
Tel. (091) 84 5125

Left: *Amidst the arid Chichester Range some cool vegetated gorges with permanent waterholes may be found.*

Nambung
National Park

One of the most extraordinary places on the coastal plains of Western Australia, the Pinnacles Desert is a geological oddity of strange, petrified twigs and limestone pillars rising up to 6 m high from the windswept desert sands. These are the calcified remnants of the root systems of trees and shrubs which once stabilised the sand dunes on the shores of the Indian Ocean, and are just a part of the area enclosed by Nambung National Park.

Facilities

There are no established camping areas in the park, although visitors can stay at the caravan and camping ground in the nearby town of Cervantes where all facilities are provided (tel. 096 52 7060). Camping is not permitted in the park.

Rental accommodation is available in Cervantes between June and October only. Picnic tables and gas barbecues are provided at some of the coastal beach areas. Wood fires are not permitted.

Location and Access

The park is 244 km north of Perth via the Brand Highway and sealed Bibby Road towards Cervantes. Turn off before Cervantes along the unsealed Grey Road into the park.

Roads within the park are unsealed and although rough in some sections they are negotiable by conventional vehicles.

The Pinnacles Desert in Nambung National Park is studded by limestone pillars which formed around 30 000 years ago.

The road around the Pinnacles Desert is one-way and must be negotiated with care, however upgrading of this and other tracks is underway.

Tours of the Pinnacles Desert are run from Cervantes and Perth.

Landscape

The park covers an area of 17 487 ha of mainly sand dunes covered in parts by low, open woodlands. The Pinnacles Desert is the most striking feature of the park, being the result of many thousands of years of limestone deposition in channels left in the sand dunes by roots of earlier vegetation. In recent times the sand has blown away from these hardened limestone pillars, leaving them standing as grey monoliths and slender limestone twigs above the yellow, quartz soil.

The Nambung River flows through the park after the winter rains, leaving little more than swamp and temporary lakes in the summer. The coastline boasts magnificent long beaches backed by heaths and thick scrub in places.

Flora and Fauna

Emus and western grey kangaroos are commonly encountered, brush wallabies are often seen at the eastern side of the park, there are many reptiles, including bob-tailed lizards, goannas and a number of snakes.

More than 90 bird species have been recorded in the park. Nankeen kestrels can be seen hovering above the heathlands, while swans and ducks use the lakes. In the thick scrub are white-winged wrens and mallee fowl.

The flora is associated with the system of sand dunes running parallel to the coast. The foredunes support such species as the pig-face, the blue-flowered *Scaevola Australia* and spinifex grasses. Coastal scrubland behind the foredunes consists of such species as the grey-leaved *Olearia axillaris*, acacias and the parasitic dodder plants. Further inland are woodlands of the tuart, *Eucalyptus gomphocephala*, banksias and casuarinas.

Walking Tracks

A number of tracks lead to beaches and picnic areas from the coast road. The best way to investigate the Pinnacles Desert is on foot, although there are no marked tracks.

Other Activities

There are many good swimming and fishing areas on the coast, and excellent opportunities for photography.

Climate

The summers are hot and dry, with modifying sea breezes. Most of the annual rainfall of 600 mm falls in the winter months which are generally mild. Visit in the springtime for wildflower displays, otherwise any time of year.

Bookings and Fees

Fees not payable for park entrance.

Information

Nambung National Park
Lot 182, Drummond Way
Cervantes Western Australia 6511
Tel. (096) 52 7043

Showy giant banksia (Banksia grandis), *one of the many beautiful wildflower species in the Stirling Range National Park.*

Stirling Range
National Park

In the heart of the great agricultural area of south-west Australia, the rugged peaks of the Stirling Range rise abruptly above the green countryside. Preserved since early this century, the whole Stirling Range has escaped the impact of farming and settlement, and the National Park is a sanctuary for an astonishing variety of flowering plants with wildflower displays unequalled in Australia.

Facilities

Caravan and car camping is allowed only at Moingup Springs near the Toolbrunup turnoff, where there are fireplaces, toilets, water, picnic tables and a ranger's residence. Next to the park entrance, opposite Bluff Knoll Road, the private Stirling Range Caravan Park has full facilities for caravanners and campers. A visitor centre is located at the park entrance.

Bush camping is allowed in the park, but the ranger should be consulted.

There are picnic areas with fireplaces at various points within the park.

Accommodation is available at the nearby towns of Albany, Mount Barker, Cranbrook, Gnowangerup and Ongerup.

Location and Access

The park is 322 km south-east of Perth via the Albany Highway, turn off on the unsealed road to Cranbrook. Albany is approximately 70 km south and Borden approximately

30 km north on the sealed Chester Pass Road which runs through the park.

Access roads within the park are mostly unsealed.

Landscape

The park covers 115 671 ha and encompasses the entire Stirling Range mountain system. A chain of rugged peaks rising to 1073 m at Bluff Knoll dominate the landscape. Ancient river systems have cut deep passes through the mountains and rolling hills which stand out starkly against the surrounding flat countryside. The higher, often mist-enshrouded peaks support a unique floral community, while the lower slopes are mainly densely covered by thick scrub.

Flora and Fauna

Western grey kangaroos, western brush wallabies and emus are occasionally seen. Quokkas, honey possums and pigmy possums inhabit the park, but are difficult to spot. Among the many reptiles are the bob-tailed lizard, the racehorse goanna, tiger snakes, dugites and death adders.

More than 100 species of birds have been recorded in the park; they include the colourful western rosella, red-capped parrot, the golden whistler and western spinebill. Wedge-tailed eagles can be seen soaring above the park.

Small forests of jarrah, marri and wandoo occupy the sandy areas between the mountains. There are many rare, unique and beautiful wildflowers in the park, including mountain bells, orchids, banksias, grevilleas and dryandras. There are stunning wildflower displays all year round, but springtime excels.

The age-old peaks of the Stirling Range rise abruptly out of the densely wooded countryside.

The peak of Mt Trio (856 m) rises majestically in the early morning light.

Walking Tracks
A number of maintained walking tracks cross the park varying from easy nature trails to strenuous climbs to the peaks. The views over the surrounding countryside are spectacular, but walkers should be prepared for sudden cold changes in winter and spring. Water should be carried in the summer months.

Other Activities
Rock climbers will find many challenges on the steep cliff faces, but the ranger should be informed beforehand. There are many opportunities for photography and nature study. Visitors are asked not to pick flowers.

Climate
Winters are cool with occasional snowfalls on the peaks. Summers are dry and warm to hot. Visit in spring for the most spectacular wildflower displays, otherwise all year round.

Bookings and Fees
Bookings are not accepted for the camping areas, although permits are required and fees are payable. Fees are not payable for park entrance.

Information
Ranger-in-charge
Stirling Range National Park
C/o P.O. Amelup
Via Borden
Western Australia 6338
Tel. (098) 27 9230

Torndirrup National Park

Reaching east to the Flinders Peninsula the rugged coastline at Torndirrup National Park is a maze of chasms, cliffs and blowholes. Panoramic views of the ocean, the islands in King George Sound and Albany can be obtained from Bald Head, the most easterly point of the promontory, while a 25 m viewing platform at The Gap allows visitors to observe the heavy south-western seas rushing up a deep narrow crevice between granite cliffs.

Facilities
Camping is not permitted in the park but accommodation is available at Albany, Frenchman Bay and Little Grove.

Location and Access
The park is situated 13 km south-west of Albany via Frenchman Bay Road and bitumen roads traverse the park.

Landscape
The coastline includes many interesting formations such as the Natural Bridge and The Gap, together with sandy beaches and rocky headlands. Beyond this the 3870 ha park is dominated by scrubland dotted with rocky outcrops.

Flora and Fauna
The vegetation is comprised of low heathland scrub such as the prickly parrot bush, with groves of swamp banksia, a profusion of peppermint clumps and a local bottlebrush with deep red flowers. New Holland honeyeaters are abundant and button quail, western rosellas and various kinds of seabirds are also seen. The western grey kangaroo and the short-nosed bandicoot are the most common mammals in the park.

Walking Tracks
There are easy walks along the firebreaks and a 2 km trail to the Peak Head area is well marked. A difficult 10 km return walk to Bald Head includes some steep climbs.

Other Activities
Fishing is popular. Rock fishermen should be aware of the possibility of 'king waves' which have already caused several drownings.

Climate
The park enjoys a fairly warm and sunny climate year round. From October to December when wildflowers are blooming is especially pleasant.

Left: *The deep red Albany banksia, a characteristic heathland wildflower at Torndirrup.*

Right: *Steep granite headlands cloak the rugged coastline along the south-western shores of Torndirrup.*

The Natural Bridge near Albany is one of many impressive rock formations in Torndirrup National Park.

Bookings and Fees

There are no park entrance fees or bookings required.

Information

The Ranger
Frenchman Bay Road
Little Grove
P.O. Box 904
Albany
Western Australia 6330
Tel. (098) 44 4090

Walpole-Nornalup
National Park

On the most southerly point of Western Australia the remnants of the once mighty karri forests stand majestically above beautiful bays and sheltered inlets. Protected within the boundaries of the Walpole-Nornalup National Park, these legendary forests shelter and nurture an abundant population of birds and a lush floral undergrowth. Ocean beaches, meandering rivers and large wilderness areas combine to make a visit a rewarding and unforgettable experience.

Facilities

The caravan and car camping area at Coalmine Beach, under shady peppermint trees on the northern shore of Nornalup Inlet, is one of the best in the state, with fireplaces, toilets, showers, a kiosk and ranger's residence. There are private caravan and camping areas on the edge of the park at Rest Point (tel. 098 40 1032), Peaceful Bay (tel. 098 40 1035) and Nornalup (tel. 098 40 1107) with cabin accommodation. Walpole has motel and hotel accommodation. Bookings are required during peak holiday periods.

There are boat ramps at Rest Point, Coalmine Beach, Nornalup and Peaceful Bay, and picnic areas at many locations in the park.

Location and Access

The park is 425 km south of Perth, 121 km west of Albany and 117 km south-east of Manjimup. Access is via the South Western Highway. Many of the access roads within the park are sealed.

Landscape

The park covers 18 116 ha of coastal hills supporting karri and tingle forests in the northern areas, coastal sandplains, and a large wilderness area south-west of Nornalup Inlet has heaths, scrub and low eucalypt forests. The Frankland and Deep Rivers flow through the park into Nornalup Inlet, which opens into the Southern Ocean with long beaches and many sheltered bays.

Karri forests surround the tidal waters of Nornalup Inlet in Walpole-Nornalup National Park.

Wattle gum fantasy, Walpole-Nornalup National Park.

Flora and Fauna

Western grey kangaroos, the yellow-footed marsupial mouse, pigmy and ringtail possums and bandicoots may be seen in the park, and it is thought that the long-nosed potoroo still exists in the area.

Bird life is abundant in the forests, and flitting below the high canopy are red-winged and splendid wrens, western rosellas, flocks of purple-crowned lorikeets and the golden whistler. Ospreys and fish hawks frequent the ocean shores, while around the inlets are many waders, pelicans and black swans.

The giant karri trees dominate the forests, growing up to 90 m tall, with girths of 8 m. They rise above an understorey of banksias, peppermint trees, flowering shrubs, white clematis, coral vines, karri wattles, crowea, ferns and many orchids. The western sandy country supports stunted banksias, eucalypts and peppermint trees. Tingle forests grow alongside the karri trees around the Frankland and Deep Rivers. In the east are red flowering gums on the low sand ridges.

Walking Tracks

The park has many walking tracks through forests and the open coastal country, varying from short nature trails to long hikes over heaths and through low forests to isolated coves between high headlands. The karri forest walks are an outstanding experience, rich with flora and fauna. A number of lookouts with spectacular views are easily accessible by foot.

Other Activities

Fishing in the rivers and ocean, swimming, surfing, windsurfing and boating are popular activities. There is a yacht club at Coalmine Beach.

Osprey can often be seen at Walpole-Nornalup National Park.

Climate
Most of the rainfall occurs in the cool winter months. Summers are warm. Visit all year round.

Bookings and Fees
Bookings should be made during peak periods for the caravan and camping ground at Coalmine Beach. Camping and park entrance fees are payable.

Information and Bookings
Ranger-in-charge
Department of Conservation and Land Management
C/o Post Office
Walpole
Western Australia 6398
Tel. (098) 40 1027

Caretaker
Coalmine Beach Caravan Park
Department of Conservation and Land Management
C/o Post Office
Walpole
Western Australia 6398

Right: *The graceful black swan is a popular attraction at Yanchep National Park.*

Yanchep National Park

On the coastal plains 53 km north of Perth, Yanchep National Park is best known for its numerous and beautiful limestone caves. Stately stone buildings and graceful gardens slope down to Loch McNess creating an atmosphere of established charm and ease in one of Australia's oldest National Parks. The wonderland cave world, scenic beauty and recreation facilities have made Yanchep a principal venue for relaxing outdoor activities and tourism.

Facilities
Yanchep boasts a variety of recreational facilities including a swimming pool, tennis court, nine-hole golf course and two ovals available for cricket, football and softball.

Beautiful grassed picnic areas with barbecue and toilet facilities are found along the eastern side of Loch McNess as well as in other quieter areas of the park. At the entrance to Boomerang Gorge, a picnic ground with facilities for disabled persons has been developed.

Meals and refreshments are available from McNess House or the Yanchep Inn, a licensed premises which also provides motel-style accommodation.

Right: *A beautiful study of waterlilies which can be seen in Yanchep National Park.*

Location and Access

Yanchep is accessible via the Wanneroo Road, north of Perth, which traverses the park. Sealed roads and all-weather vehicle access are available.

Landscape

The park is roughly triangular in shape, covering 2799 ha. Set in limestone hills which rise to heights of around 88 m above the beautiful Loch McNess, the caves contain interesting caverns and formations. In Crystal Cave, massed stalactites hang over the wide pool of an underground stream. The sandy soils of the coastal plains support a rich array of vegetation and wildlife.

Flora and Fauna

The bushland reserve established to preserve the wildlife and flora is dominated by massive tuart, jarrah and marri. Smaller shrubs such as banksia and casuarinas also occupy a large part of the park. A wealth of wildflowers may be seen in the spring, including red and green kangaroo paw, varieties of banksia, native buttercup and wisteria and Christmas bush. At the edge of the water and across the swampland, sword rush, and yanget or bullrush are found.

Big grey kangaroos often appear at sundown when they emerge to graze on the ovals and open lands. Brush wallaby, brush-tailed possums, bats, native rats and echidnas are also present.

Water birds include black swans, grebes, pelicans, cormorants, egrets, herons and ducks. In the varying habitats, from woodland to marsh and open water, hawks, gulls, kingfishers, parrots and many other species survive. Reptiles, frogs and tortoises are also common throughout the park.

Walking Tracks

Around Loch McNess, Yanjidi walk trail, a documented track, gives descriptions of the various animals and vegetation along the way. A number of other walking trails exist but advice on their use should be sought at the park office. Marked nature tracks are also designed for disabled persons.

Other Activities

Guided tours of the Crystal Caves highlight the beautiful formations there. These caves are also open for public inspection at various times. On Loch McNess a fleet of rowing boats is available for daily hire for tours around the waterway.

In addition, a wildflower and fauna complex displays some of Australia's unique flora and fauna including a colony of koalas. Gloucester Lodge Museum, a magnificent holiday lodge built in 1933, contains a collection of artefacts from the park and surrounding district.

Climate

Yanchep has a temperate climate similar to that of Perth with an average daily temperature of around 18°C. Visits to the park may be enjoyed year round.

Bookings and Fees

Fees are payable for park entrance, recreation facilities and cave and launch tours. The Yanchep Inn is privately managed and tariffs are available from the licensee.

Information and Bookings

Ranger-in-charge
Yanchep National Park
C/o P.O. Yanchep
Western Australia 6035
Tel. (095) 61 1661
Information Office: Tel. (095) 61 1004

Licensee
Yanchep Inn
Yanchep National Park
Tel. (095) 61 1001

Tasmania

Tasmania, the world's most mountainous island, embraces hidden treasures for its visitors: jagged peaks, limestone caves, torrential rivers, deep valleys cushioning sparkling glacial lakes, forested hillsides, gale-swept peninsulas and long sweeps of sandy beaches.

Almost seven per cent of the total area of the state has been set aside for National Parks, all of which are easily accessible from Hobart or Launceston.

Alpine scenery dominates the Tasmanian highlands in the north of the state. Here, craggy peaks and bluffs, hundreds of glittering tarns and deeply carved, rock-strewn valleys were formed about 20 000 years ago during the last ice age. Alpine heaths and gnarled, low-growing shrubs struggle to survive under the snow, but in spring the moorlands become a wilderness of flowering heaths.

The National Parks in this ice-moulded area include Tasmania's premier ski fields at Ben Lomond, and the stark, unforgettable mountain scenery of the Cradle Mountain-Lake St Clair National Park.

In contrast to the majestic severity of the highlands, the National Parks on the coast of northern Tasmania enjoy a milder climate; here is Freycinet National Park with its long white beaches framed by granite cliffs, and Mount William National Park with its coastal heathland and dry eucalypt forest providing sanctuary for an abundance of wildlife.

South-western Tasmania is dominated by water, with its weeks of mist, rain and snow, its sodden buttongrass plains and its clear ice-cold streams. Here, Southwest and Wild Rivers National Parks constitute true wilderness areas, providing Australia's best rafting areas and most celebrated system of mountain-and-ocean walking trails.

Closer to Hobart, Mount Field National Park is an astonishing kingdom of trees; and the parks and reserves to the east of Hobart offer arresting coastal scenery, and fascinating limestone caves.

Ben Lomond National Park

Ben Lomond's lofty plateau, with its precipitous sides, is the highest region of north-eastern Tasmania and the state's premier snow sports area. Deep beds of dolerite, smashed and scoured by the movements of glaciers in the last ice age, have formed the imposing cliffs at the plateau's edge. The park offers adventure to those walkers and climbers hardy enough to brave the chilly climate; in season the snow-covered slopes provide a fairyland atmosphere for the many skiers who flock here and, out of season, botanists and geologists find much to interest them.

The elusive platypus has been seen on the plateau in Ben Lomond National Park.

Facilities
There is a tavern and car park at the ski village, about 2 km from the top of Jacob's Ladder. A public toilet is nearby and there is a day shelter hut with gas heaters and a gas cooking stove on the ski fields. This is in operation during winter only. The downhill slopes are serviced by seven tows, and ski schools and a ski hire business, which also sells food and drink, operate from the village. The lodges are all owned privately or by alpine clubs.

Cross-country skiing is popular on the mountain but care should be taken in case of bad weather as only routes close to the village are marked by poles. Both cross-country skiers and walkers are advised to register details of their trips with the ranger.

Camping is only permitted by those undertaking extended walks and there is no firewood available on the plateau.

Location and Access
Ben Lomond National Park is located 50 km south-east of Launceston. Access from Launceston is via White Hills; from Hobart via Evandale and White Hills on to the Blessington road. Turn right on to the Ben Lomond road 3.5 km before reaching Upper Blessington. The road to the plateau is unsealed and during adverse weather conditions it may be closed without notice.

Landscape
The park covers a wide, flat-topped dolerite plateau, the highest part of which is Legges Tor. The plateau is roughly 14 km in length by 6 km in width and is in excess of 1300 m in height. During the ice age about 20 000 years ago, the plateau top was covered by ice, and today shows evidence of this in its extensive areas of bare rock. Around the precipitous edges the rock has been shattered by frost and the masses of huge boulders form extensive blockfields on the slopes.

Flora and Fauna
Slow-growing, short and often stunted plants with small hard leaves grow on the exposed mountain tops, however, the alpine moors are fertile with several species of cushion plants, some endemic to Tasmania, appearing through the melting snow, and heaths flower brilliantly in summer. Celery top pine and stunted, deciduous beech called tanglefoot—both peculiar to Tasmania—grow at subalpine levels, above eucalypt woodlands.

Several native mammals make their home on the Ben Lomond plateau. The wombat, Tasmanian devil, echidna, pademelon, Bennetts wallaby and the brush-tailed, ringtail, sugar glider and pygmy possum can survive the bleak conditions. Even a platypus has been observed mid-winter on the plateau. Native rats, such as the velvet-furred rat, have successfully adapted to this alpine area, forming extensive systems of runways beneath the vegetation

Several birds are found here including wedge-tailed eagles, hawks, crows, currawongs and honeyeaters, but the meadow lark is the only real resident of the plateau.

Walking Tracks
There are few marked walking tracks on the plateau, although some routes are poled or cairned. A track extends from Carr Villa on the slopes of the mountain to Legges Tor, the second highest peak in the state. Day visitors are advised to be off the mountain before 4 p.m. as very rapid weather changes occur without warning. Walkers are advised to register their trip with the ranger.

Other Activities
Although Ben Lomond is considered to be mainly a skiing area, walking, nature study, rock climbing and photography may be readily enjoyed in the park.

Climate
In any season weather conditions may be severe and it is essential that warm clothing with waterproof over-clothes be worn in winter. Visit during the winter months for the snow.

Bookings and Fees
Bookings and fees are not required.

Information
The Ranger
Ben Lomond National Park
C/o P.O. Box 180
South Launceston
Tasmania 7249
Tel. (003) 41 5312

Cradle Mountain-Lake St Clair

National Park

Encompassing 131 915 ha of stark unforgettable mountain peaks, lakes and alpine moorland, Cradle Mountain-Lake St Clair National Park is Tasmania's most famous park. The park extends from Pencil Pine Creek near Cradle Mountain in the north over rugged dolomite-capped ranges including Mount Ossa (1617 m), The highest mountain in the state, down to Derwent Bridge, south of Lake St Clair, a spectacular glacial lake with a depth of over 200 m. The park contains unsurpassed walking country, representative of Tasmania's best wild alpine and mountain wilderness landscape.

Facilities

Cradle Mountain-Lake St Clair is administered in two sections with rangers located at Cradle Mountain and at Cynthia Bay near Derwent Bridge. Food and petrol supplies and accommodation are available at the northern park entrance, 2 km from the new camping ground at Cradle Mountain. There are a number of huts with gas stoves, cooking utensils and mattresses. Light is supplied to all huts. Amenities blocks include hot showers and toilets.

A picnic area with barbecues, toilets, a day-use hut, kiosk, boat ramp and boat service is located at Cynthia Bay. Overnight huts are also available with beds, heating and tap water. Amenities blocks contain showers, toilets and laundry facilities. Caravan and tent sites are provided, some with power. Derwent Bridge has a hotel and a petrol station with a café.

Bush campers must notify the ranger of their itinerary and check out with him or her on completion of their basic walk. There is a network of basic huts within the park about every 10 km or so but it is essential that a tent and stove with enough fuel for the whole trip be carried as open fires are banned on the track.

Location and Access

Cradle Valley is located 85 km from Devonport via Forth, the last 10 km of which is a narrow and unsealed road. Lake St Clair is 173 km from Hobart via the Lyell Highway and about the same distance from Launceston via Deloraine or Poatina

The brilliant glacial waters of Lake St Clair provide opportunities for fishing, sightseeing and sailing.

and the Lyell Highway. At Derwent Bridge a 5.5 km park access road is subject to occasional snow blockages in winter.

Bus services operate along the Lyell Highway between Hobart and Queenstown, and from Launceston to Cradle Mountain.

Landscape

In the north, the landscape is dominated by Cradle Mountain and Barn Bluff. High tranquil lakes are found amidst steep crags and deep gorges while open moors and heathlands are surrounded by towering rock walls.

The present landforms are largely the outcome of glacial activity. Lake St Clair, with an area of 22 sq km, was gouged by ice activity extending down the valleys from the north. Moraine ridges, formed from glacial debris, extend south from the lake. They are up to 9 m high and 3 km long. Because they interfere with local drainage, the terrain has alternating areas of boggy land and well-drained ridges.

Flora and Fauna

Around Cradle Mountain and Lake Dove there are dense forests of deciduous beech, Tasmanian myrtle, pencil and King Billy pine with undergrowth of mosses and ferns. The open heathland down through Cradle Valley features a wealth of spring and summer wildflowers.

Above: *Brush-tailed possums may sometimes be seen at night in Cradle Mountain-Lake St Clair National Park.*

Left: *The snow-covered peak of Cradle Mountain in Cradle Mountain-Lake St Clair National Park.*

Chilly mountain-fed streams in Cradle Mountain-Lake St Clair National Park are a home for platypuses and native trout.

Moving towards Lake St Clair many different plant communities exist including eucalypt forests on the drier mountain sides and buttongrass sedgelands in the waterlogged lowlands. Banksias flower year round and Tamania's climbing heath displays crimson bells in early autumn.

Native mammals such as the Bennetts wallaby and pademelon are often seen while abundant nocturnal animals include brush-tail possums, native cats and the Tasmanian devil. Platypus exist in the mountain lakes and streams which are also inhabited by native trout.

Bird life includes the black currawong and the flightless native hen among many other species.

Walking Tracks

The Overland Track is an 85 km walking tail linking the two main access points in the park. The track is relatively easy and safe for active persons providing suitable equipment and adequate precautions are taken. The rangers should be consulted and walks must be registered. A fee is payable at the office if you are walking this track.

Other Activities

In season, licensed trout fishing is permitted in the lakes and rivers. Private boats may be used for fishing or scenic trips. Swimming is permitted but the lakes are fed by mountain streams and are always cold.

Rock, snow and ice climbing are popular while cross-country skiing is enjoyed by experienced skiers when conditions are suitable.

Climate
The weather can be unpredictable and severe even in summer. Rain, sleet and snow occur in all seasons. The southern end of the park is more temperate but still very changeable so visitors should be prepared with woollen and waterproof clothing at all times.

Bookings and Fees
Entrance and camping fees are payable at Cradle Mountain and Lake St Clair. Accommodation is heavily booked especially during school and public holidays. Advance bookings are preferable, taken by the ranger-in-charge at Lake St Clair and by Cradle Mountain Lodge at the northern end of the park.

Information and Bookings
Ranger-in-charge
National Parks and Wildlife Service
Cradle Mountain
Tasmania 7306
Tel. (003) 63 5187 (for information)
Tel. (003) 63 5164 (for bookings)

Ranger-in-charge
National Parks and Wildlife Service
Lake St Clair
C/o P.O. Derwent Bridge
Tasmania 7465
Tel. (002) 89 1115

Freycinet National Park

Precipitous headlands broken by deep indentations and sandy beaches form the shores of Freycinet National Park, an area on Tasmania's east coast which offers over 10 000 ha including both Freycinet Peninsula and Schouten Island. The magnificent coastline vista comprises red granite summits, off-shore islands, lagoons and picturesque bays, while coastal scrubs and heaths flowering profusely in spring contribute to the scenic beauty of the region.

Facilities
A camping area with some powered sites is situated at the park entrance. Other camping areas are available for bushwalkers at Wineglass Bay, Hazards Beach, Cooks Beach and Bryans Beach. There is also a camping area at Morey's Bay on Schouten Island. The ranger must be consulted for site allocation in the camping area at the park entrance.

A privately owned 'chateau' at Coles Bay offers hotel-style accommodation and a small youth hostel is sited near The Fisheries. Picnic, barbecue and toilet facilities are available and supplies may be purchased from the Coles Bay township. Drinking water is sometimes scarce so walkers should carry a supply.

Location and Access
The park is situated 60 km north of Swansea via a road leading from the Tasman Highway. The turnoff is well signposted 12 km south of Bicheno and 32 km north of Swansea. The road into the park is being sealed but 8 km remains gravel.

Landscape
Freycinet Peninsula has a high central spine dominated by Mount Freycinet (620 m). To the north, the steep four-peaked ridge of the Hazards juts into a marshy isthmus bordered by Hazards Beach and Wineglass Bay.

Schouten Island, lying to the south, is an uninhabited island accessible by boat, with landscape ranging from sandy beaches to high rocky cliffs. It is split by a clear faultline, the north side consisting mostly of bare granite while the timbered heights are based on dolerite.

Flora and Fauna
Despite extensive fires which swept through the park in 1980 and 1981 much of the vegetation has recovered well. Eucalypts form the bulk of tree cover with patches of she-oak, banksia and wattle. In the spring, coastal heathlands provide a colourful display of wildflowers including boronia, wedding bush and kangaroo tail.

Left: *The safe, sheltered swimming beach of Honeymoon Bay viewed from Mt Amos in Freycinet National Park.*

Right: *Ruins of a nineteenth century church at Port Arthur on the Tasman Peninsula.*

Picturesque tranquillity and seclusion of Wineglass Bay in Freycinet National Park.

Brush possums, native cats, wombats and wallabies are some of the most common mammals. The bird life, too, is prolific. Forest birds include the butcherbird, green rosella, black cockatoo and yellow wattlebird. Many sea birds such as gulls, fairy penguins and gannets may be seen.

Walking Tracks

There are a number of excellent walking tracks over or around the Hazards peaks and extending to Bryans Beach in the south. The ranger should be informed by bushwalkers intending to make overnight treks.

Other Activities

Water sports include swimming, sailing, canoeing, water-skiing and skindiving. The eastern side of the park, which is exposed to Tasman Sea swells, can be dangerous but sandy beaches on the other side are quite safe. Off-shore fishing grounds are found around the peninsula and Schouten Island. Two boat launching ramps are located in Coles Bay and a charter boat service operates for Schouten Island or fishing trips.

Climate

Winter is mild and sunny with little wind and is generally a more pleasant time to visit the park. Spring and summer can bring unsettled weather.

Bookings and Fees

Fees are payable for camping, at the park entrance only. Demand for campsites is heavy, especially during holiday seasons, so early booking is recommended.

Information and Bookings

The Ranger
Freycinet National Park
Via Coles Bay
Tasmania 7215
Tel. (002) 57 0107

Maria Island

National Park

Off the eastern coast of Tasmania, seclusion and tranquil charm may be found amidst the scenic grandeur of Maria Island National Park. Linked by a low narrow isthmus lined with sandy beaches, two steep, forested, land masses provide visitors with opportunities for both peaceful bushwalks and challenging rock climbs. Interesting ruins can be explored at the historic township of Darlington on the northern end of the island.

Facilities

A camping site as well as gas and wood barbecues, toilets and wash troughs are found at Darlington. Hot water and showers are not available and there are no shops, so visitors must bring all supplies with them.

The Penitentiary at Darlington offers basic accommodation in the form of units with bunks and a pot-belly stove.

At Frenchs Farm and Encampment Cove, basic camping areas exist with pit toilets and communal fireplaces but visitors should carry portable stoves and water to Encampment Cove.

There is a visitor centre near the wharf which can provide information regarding facilities and activities. A public telephone is also available.

Location and Access

Maria Island is situated 88 km north-east of Hobart, approximately 1 1/2 hours drive along the Tasman Highway via Orford to the Louisville Resort. A ferry runs daily between the wharf at Louisville and Maria Island. Departure depends on passenger numbers and weather conditions. Chartered aircraft are also available.

Cape Barren geese thrive on the lush grassy areas of Maria Island.

Landscape

Maria Island comprises approximately 10 000 ha, the seaward side of which is dominated by precipitous sandstone and limestone cliffs while the inner side slopes gently to the sea. Swamps and lagoons are dotted along the foreshores. The northern and southern ends of the island were split by major faults over 40 million years ago when the land was connected to mainland Tasmania. More recently the sea flooded in, forming Mercury Passage which separates Maria Island from Tasmania. McRaes Isthmus connecting the two halves of Maria Island is a thin bar formed by the deposition of sand by the waves.

Flora and Fauna

Most of the slopes of the island are covered by dry eucalypt forests but there are also areas of open heathland and ferny creek gullies. On some of the higher ridges, small areas of rainforest are to be found. Europeans have introduced species such as radiata pine, canary broom and thistles. Along the coast, sand dunes have a cover of introduced marram grass as well as spinifex and low-growing shrubs and succulents.

In 1965, the Tasmanian Government took over the island as a fauna reserve and began introducing animals whose habitat on the mainland was threatened. These include Forester kangaroos, Bennetts wallabies, emus, possums, bandicoots and echidnas. Cape Barren geese also thrive on the grassy areas.

Maria Island has an abundant variety of bird life. Noted among these is the endangered forty-spotted pardalote. Marine animals such as bright red sea-anemones, sea stars and crabs, as well as rock lobsters, abalone and many species of fish, dwell in the rockpools and waters around the island.

Walking Tracks

There are a number of walking trails around the coast and through the bushland. These are signposted and information regarding them may be obtained from the ranger. A special one to one-and-a-half hour walk along Fossil Cliffs points out obvious features of interest. A map and notes are available from the visitors centre.

Other Activities

Maria Island has many safe swimming beaches and surfing is possible at Riedle Bay. Skindiving and scuba diving are popular and boats can be chartered from Orford and Triabunna. Those venturing out in boats should note that weather conditions are subject to sudden changes.

Climate

The island enjoys a generally pleasant mild climate but, like all of Tasmania, quite cold changes are possible at any time of the year.

Bookings and Fees

No park entrance fees are payable, although there is a small charge for camping and accommodation at the Penitentiary. Bookings should be made in advance especially in summer when the island is very popular with tourists. Seats on the ferry *James McCabe* from Louisville should also be booked.

Information and Bookings

The Ranger
Maria Island
C/o P.O. Triabunna
Tasmania 7190
Tel. (002) 57 1420

M.V. *James McCabe*
K. Woodward
Tel. (002) 57 1362

Bennetts wallabies, native to Tasmania, are often seen in Mt Field National Park.

Mount Field National Park

Tasmania's first National Park, Mount Field, is a region of astonishing forests, waterfalls, wild moorland and rugged mountain ranges. Spectacular Russell Falls plunges in two stages into cool shady gorges thick with treeferns. This well-known attraction, together with the developed ski fields within the park and the magnificent mountain and forest scenery, help make Mount Field, only a one-hour drive from Hobart, one of the most popular parks in the state.

Facilities

Near the park entrance there is a camping area and caravan park with toilets, hot water, showers and laundry facilities. Picnic shelters, electric barbecues and fireplaces are also provided.

There is a kiosk near the caravan park selling basic foodstuffs and souvenirs and a visitor centre providing information about the park.

At Lake Dobson, 16 km into the park, there is a picnic shelter and toilets. Three accommodation huts here provide basic facilities, and bookings may be made with the ranger. Thirty minutes' walk from the lake, at the ski slopes, there are four ski tows, a kiosk, toilets and a public shelter as well as ski hire and instruction if required. These facilities operate only during winter. Accommodation at club lodges on the slopes is generally restricted to members and their guests.

Location and Access

Mount Field is reached on a sealed road from Hobart via Gordon River Road. The road to Lake Dobson may sometimes be closed for short periods during winter, and wheel chains should be carried.

Landscape

Mount Field National Park covers an area of 16 257 ha. From the dense eucalypt forests of the lowland regions, the terrain ranges through rainforest gullies with many beautiful waterfalls, to high bare ridges and moorlands beyond Lake Dobson, culminating in Mount Field West (1434 m). There are numerous lakes and mountain tarns scattered across the plateau.

Flora and Fauna

The entrance to the park is guarded by a forest of towering swamp gums or mountain ash. Beneath these, native musk and myrtle form a mixed undergrowth. Canopies of manferns together with many other fern species grow around the waterfalls and other waterways.

As the terrain rises, rainforest communities of myrtle, sassafras, native laurel and celery top pine are to be found while higher still more hardy trees such as snow gums and yellow gums exist. Around Lake Fenton, King Billy pine, pencil pines and thickets of fagus are found. The mountain tops and plateaux support tough weather-resistant plants such as alpine conifers, pineapple grass bogs, herbfield and cushion plants.

There are over 50 identified bird species at Mount Field. The green rosella, black currawong, Tasmanian thornbill and scrub tit are quite common while the lyrebird, an introduced species, flourishes in the rainforest areas and fern gullies.

Pademelons, Bennetts wallabies and the barred bandicoot are a frequent sight, grazing around picnic areas at dusk. The wombat and the Tasmanian devil live in the upper areas while native cats range over most of the park. Platypus can be seen in many mountain lakes and in the Tyenna River bordering the park. In highland pools, the mountain shrimp, an ancient species, may be found.

Walking Tracks

There are a variety of clear, well-marked tracks offering walks ranging from an easy stroll to a strenuous hike. Brochures are available for a number of these including the disabled persons' trail to Russell Falls. These can be obtained from the kiosk or park headquarters.

Magnificent Russell Falls plummets into a gully of tree ferns in Mt Field National Park.

Other Activities
A visit to Russell Falls is highly recommended with the walk taking a leisurely 30 minutes from the carpark near the park entrance.

Skiing is especially popular in winter with runs varying from beginners to advanced. In good seasons the highland area provides excellent opportunities for cross-country skiing. A small museum showing early skiing in the park is located at the Twilight Tarn hut.

Climate
The weather is unpredictable and sudden changes can occur without warning. It is essential that alpine walkers are well equipped. Rainfall is high throughout the year and bitterly cold weather is to be expected during winter.

Lyrebirds flourish in the rainforest areas of Mt Field National Park.

Bookings and Fees
Park entrance, camping and hut accommodation fees are payable. Bookings are only accepted for Christmas and Easter and further information is available from the ranger at the park headquarters.

Information and Bookings
The Ranger
Mount Field National Park
P.O. Box 41
Westerway
Tasmania 7140
Tel. (002) 88 1149

Mount William National Park

In the far north-eastern corner of Tasmania, Mount William National Park is an area of granite headlands, long sweeps of white beaches, coastal plains and ranges. The park is primarily a wildlife reserve—the area providing a major sanctuary for Tasmania's only kangaroo, the Forester, and several offshore islands providing sanctuary for seabirds.

Facilities
Camping areas have been developed at several sites along the coast road leading south from Musselroe Road, and along the road from Eddystone Point to Picnic Rocks. All campsites are marked with signs, and toilets, fireplaces and picnic tables are provided but there is no permanent drinking water. There are launching sites for small craft near most of the campsites.

Location and Access
The park is located 30 km from St Helens; 8 km from Gladstone. To reach the northern end of the park follow the road from Gladstone towards Musselroe Bay—two roads enter the park from this point. The road to Eddystone Point provides access to the southern section of the park, a branch road runs north for about 4 km to Picnic Rocks.

Landscape
Mount William National Park covers 13 812 ha between Ansons Bay and Musselroe Bay. The bays and granite headlands are a feature of the park, but it also includes flat coastal plains which contrast with Mount William (216 m) and Bayleys Hill (184 m).

To the east, about 3 km from the coast, are Georges Rocks, a low-lying group of small rocky islands that provide sanctuary for several species of seabirds which nest here.

The coastline is strewn with submerged rocks and reefs and the lighthouse at Eddystone Point was built in 1889 to prevent the many shipwrecks in the area.

Rich in native fauna, the Mount William area was probably an important hunting area for the Tasmanian Aborigines. Today, the main signs of their presence are the piles of discarded shells along the dunes, known as kitchen middens.

Flora and Fauna
An extensive area of coastal heathlands produce a spectacular display of wildflowers in spring and early summer when the red and white heaths provide a beautiful background for the contrasting colours of the golden wattles and guinea flowers.

The park also contains the largest area of coastal dry sclerophyll forest in the state. The dominant trees being black peppermint with a small area of black and white gum. Banksia, she-oaks, bull-oaks and grass trees are also found.

Sand dunes back the coastline and here succulent creeping plants and specialised grasses grow. Behind the dunes are a number of temporary marshes and paperbark swamps.

The park provides an important sanctuary for the Forester kangaroo, and large areas of pasture are being maintained as

The Tasmanian Devil is a common sight in Mount William National Park.

grazing land for them. Other animals commonly found here include the Bennetts wallaby, pademelon, echidna, wombat, brush possum, native cat and Tasmanian devil.

Bird life is rich and varied with over 100 recorded species. Honeyeaters, blue wrens, several species of robin and the fire-tail finch inhabit the heaths and further inland the striated pardalote and kookaburra can be seen. Gulls, terns, gannets, albatrosses, the pied and sooty oystercatchers and the white-breasted sea eagle are found on the coastal area.

Walking Tracks
The coast, with its wide sandy beaches and rocky headlands, offers plenty of scenery for pleasant easy walks. There is a walking track leading to the top of Mount William.

Other Activities
Boating, fishing, birdwatching and photography are all popular activities within the park.

Climate
Visit all year round. The spring and summer flora is popular with photographers.

Bookings and Fees
There are no fees and bookings are not necessary.

Information
The Ranger
Mount William National Park
P.O. Box 1
Gladstone
Tasmania 7254
Tel. (003) 57 2108

Southwest
National Park

The scenic grandeur of magnificent mountain, bush and coastal country in the Southwest National Park is spread over the largest wilderness area in Tasmania, 442 240 ha. Embracing the lucid waters of Lake Pedder, Southwest reaches out beyond the Frankland and Arthur Ranges to the west and south coasts of Tasmania through remote, densely forested valleys and precipitous rocky ridges. Southwest, Cradle Mountain-Lake St Clair and the Franklin-Lower Gordon Wild Rivers National Parks have been classified by UNESCO as World Heritage areas and comprise one of the last great cool temperate wilderness areas in the world.

Facilities
Around Lake Pedder there are three camping areas situated at Edgar Dam, Huon and Wedge River. All provide tank water and toilets. Sheltered campsites are available at Huon, while Edgar Dam is an exposed site although it does have a shelter hut and a boat ramp.

Large areas of Antarctic beech and mountain ash form part of the massive wilderness of Southwest National Park.

Extensive buttongrass plains on the lowlands of Southwest National Park.

There are picnic areas at the Needle, Wedge River, Condominium Creek, Teds Beach and Serpentine Dam and boat ramps are provided at the lakeside sites.

Beyond the park on the northern point of Lake Pedder, the township of Strathgordon has petrol and shopping facilities, a caravan park with powered sites and the Lake Pedder Chalet which provides hotel-style accommodation. Tourist and fishing boats are available.

Elsewhere in the park shelters exist but they are few and far between. Bushwalkers should regard these as emergency shelters only. The remote and rugged nature of the area makes it necessary for precautions to be taken and the service ranger at Mount Field should be notified before overnight trips.

Location and Access

Southwest National Park occupies almost the entire southwest corner of Tasmania. The Lake Pedder impoundment is accessible by road from Hobart travelling via the Lyell Highway and turning west, north of New Norfolk via Maydena to the Scotts Peak area or to Strathgordon.

Left: *Lake Pedder, viewed from Strathgordon to the north of Southwest National Park.*

For the more adventurous there are walking tracks from Port Davey and the south coast. In addition, driving south from Hobart via Huonville and Geeveston, the road finishes in a track which leads bushwalkers to Federation Peak inside the wilderness area.

Landscape

Southwest contains a number of major mountain ranges, over 50 lakes and several river systems. Lake Pedder, a clear glacial lake nestling among jagged peaks, was submerged in 1972 by a dam built to increase the hydro-electric potential of the Gordon River.

The gouging effect of glaciation has left a landscape of sharp ridges and steep valleys. The snow-covered ranges include Precipitous Bluff and Federation Peak in the south, the rugged series of peaks known as the Arthur Ranges, Mount Anne and Scotts Peak on the eastern side of Lake Pedder and the Frankland Range.

Flora and Fauna

Towards the east of the park, many valleys have a dense cover of eucalypt forest. Endemic trees include the Huon pine and the large King Billy pine as well as Tasmanian myrtle. On the acidic, less fertile soils there are extensive areas of buttongrass plains while in the more fertile areas of high rainfall, dense temperate rainforests flourish. Antarctic beech and mountain ash upon which hang wispy golden mosses, lichens and fungi, provide strong evidence that Australia was once joined to Antarctica. The flowers of the leatherwood tree produce a particularly aromatic honey.

Many animals live in the variety of habitats within Southwest National Park including Bennetts wallabies, pademelons and wombats. Bird life is abundant; honeyeaters, brown scrub wrens and grey fantails are quite common while the white cockatoo and black currawong inhabit the mountainous regions. Several species of lizards as well as tiger, copperhead and white-lipped whip snakes are frequent.

Walking Tracks

There are a number of trails around Lake Pedder which offer pleasant, easy bushwalking. Elsewhere, however, most walking routes cross some of the ranges in the park, which average from 1000–1450 m in height.

Other Activities

Lake Pedder is popular for water sports including fishing and boating. Swimming is possible but the water is usually quite cold. The park contains some of the finest bushwalking country in Australia, however overnight walking parties must carry full equipment.

Climate

Southwest is unpredictable. Frequent storms can bring snow and freezing temperatures even in summer.

Bookings and Fees

Fees and bookings are not necessary.

Information and Bookings

Ranger's Office
National Parks and Wildlife Service
P.O. Box 41
Westerway
Tasmania 7140
Tel. (002) 88 1283

Visitor's Guide

New South Wales

Apsley Gorge National Park—6670 ha, 497 km north of Sydney. Car access off Oxley Highway east of Walcha. Spectacular gorges and waterfalls. Walking, swimming, picnicking, camping. Tel. (967) 72 1733.

Bald Rock National Park—2776 ha, 29 km north-east of Tenterfield on the New South Wales/Queensland border. Car access off New England Highway and Mt Lindesay Road. Bald Rock is a large granite dome surrounded by heath and forestland. Tel. (067) 32 1177.

Boonoo Boonoo National Park—2442 ha, 35 km north of Tenterfield. Car access by 14 km of rough gravel road off Mt Lindesay Road. Scenic river and rainforested gorge. Tel. (067) 32 1177.

Border Ranges National Park—30 800 ha, 50 km south-east of Murwillumbah. Car access off Mt Lindesay Highway. Tweed Range Scenic Drive crosses park. Variety of rainforest types, waterfalls, mountain parks, gorges, walking trails, picnic grounds. Tel. (066) 28 1177.

Brisbane Waters National Park—11 317 ha, 60 km north of Sydney. Car access via Pacific Highway. Aboriginal rock engravings, woodlands and swamps. Picnicking, fishing, bushwalking. Tel. (043) 24 4911.

Broadwater National Park—3691 ha, 759 km north of Sydney, 3 km north of Evans Head. Car access from Pacific Highway. Ocean beaches, wildflowers, fishing, swimming, bushwalking. Tel. (066) 28 1177.

Bundjalung National Park—17 544 ha, 725 km north of Sydney. Car access off Pacific Highway. Long coastline, Esk River, freshwater lagoons, coastal scenery, water sports. Tel. (066) 28 1177.

Cathedral Rock National Park—5194 ha, 609 km east of Armidale. Car access from Armidale-Ebor and Guyra-Ebor Road. Interesting rock formations, Round Mountain, forests and swamps. Tel. (066) 57 2309.

Crowdy Bay National Park—7237 ha, 45 km north-east of Taree. Car access from Pacific Highway via Coralville or Laurieton. Beaches, heath wildflowers, birds and other fauna, water sports, bushwalking. Tel. (065) 83 5518.

Dharug National Park—14 785 ha, 75 km north-west of Sydney. Car access through Wisemans Ferry or from Gosford. Hawkesbury River, abundant fauna and flora, bushwalking, camping, etc. Tel. (043) 24 4911.

Dorrigo National Park—7819 ha, 590 km north of Sydney. Car access via Bellingen-Dorrigo Road. Rainforest area with great diversity of fauna, bushwalking, etc. Tel. (066) 57 2309.

Goulburn River National Park—57 083 ha, 399 km north-west of Sydney. Difficult access via Merriwa-Mudgee Road, or through private property. Semi-arid landscape, basalt capped ridges. Tel. (065) 43 2274.

Guy Fawkes River National Park—33 854 ha, 110 km south-west of Grafton. Car access from Grafton/Armidale Road at Hernani and Dundurrabin. Rugged river wilderness rapids, gorges, mobs of kangaroos, Ebor Falls lookout. Tel. (066) 57 2309.

Hat Head National Park—6194 ha, 20 km east of Kempsey. Car access from Pacific Highway at Kempsey. Extensive coastline, surfing, fishing, bush camping. Tel. (065) 83 5518.

Heathcote National Park—2251 ha, 32 km south of Sydney. Car access via Princes Highway south of Waterfall along Woronora Dam Road. Woodland vegetation, wildflowers, swimming, camping, bushwalking. Tel. (02) 521 2230.

Kanangra-Boyd National Park—68 276 ha, 180 km west of Sydney. Car access via Great Western Highway and Jenolan Caves. Spectacular views of rugged wilderness, waterfalls, narrow canyons, limestone caves, trout fishing and bushwalking. Tel. (047) 87 8877.

Macquarie Pass National Park—1064 ha, 130 km south of Sydney. Car access off Illawarra Highway. Macquarie Rivulet gorge, dense forests, diverse bird and mammal population, bushwalking, picnicking. Tel. (044) 21 9969.

Mallee Cliffs National Park—57 969 ha, 30 km east of Mildura. No public access. Semi-arid mallee vegetation, endangered mallee fowl. Tel. (050) 23 1278.

Marramarra National Park—11 539 ha, 40 km north of Sydney. Car access from Old Northern Road or Wisemans Ferry. Mangroves, Marramarra Creek, views of Hawkesbury River, flora and fauna, canoeing, etc. Tel. (02) 457 9322.

Mootwingee National Park—68 912 ha, 132 km north-east of Broken Hill. Car access by dirt road from Broken Hill, Tibooburra and White Cliffs. Semi-arid sandstone range with waterholes, narrow gorges, Aboriginal relics. Tel. (080) 88 0253.

Mt Imlay National Park—3764 ha, 496 km south of Sydney. Access on foot from Burrawong or Imlay Forest Roads west of Princes Highway. Heavy forest, wide coastal views, rough bushwalking and camping. Tel. (0649) 61 434.

Mt Warning—2210 ha, 16 km south-west of Murwillumbah. Car access via Murwillumbah-Kyogle Road through Dum Dum. Dominated by steep jagged peak of Mt Warning surrounded by rainforest, woodlands and heath. Tel. (066) 28 1177.

Nightcap National Park—4945 ha, 35 km north of Lismore. Car access via Pacific Highway turning off at Lismore. Spectacular waterfalls, outlooks, rainforest covered ranges, walking tracks. Tel. (066) 28 1177.

Nymboida National Park—1368 ha, 65 km west of Grafton. Access on foot or by canoe. Four-wheel-drive recommended from Grafton/Glen Innes Road. Two spectacular wild rivers, canoeing, rafting. Tel. (066) 42 0613.

Thirlmere Lakes National Park—600 ha, 90 km south of Sydney. Car access south of Picton off Remembrance Driveway. Freshwater lakes, forested hills, boating, swimming and canoeing. Tel. (02) 521 2230.

Tomaree National Park—800 ha, 225 km north-east of Sydney. Car access off Pacific Highway from Raymond Terrace. Broken coastal strip, high sea cliffs. Tel. (049) 87 3108.

Warrabah National Park—3400 ha, 80 km north of Tamworth. Car access north of Manilla. Cascades, rapids and pools of Namoi River, surrounding dense vegetation and abundant wildlife, water sports and bushwalking. Tel. (067) 72 1733.

Washpool National Park—30 000 ha, 75 km east of Glen Innes. Car access off Gwydir Highway. Temperate rainforests, creeks, walking trails. Tel. (067) 32 1177.

Werrikimbe National Park—33 000 ha, 127 km west of Wauchope. Car access off Oxley Highway. Unspoilt rainforest, deep gorges, waterfalls, abundant wildlife, bushwalking and camping. Tel. (067) 72 1733.

Woko National Park—4500 ha, 350 km north of Sydney. Car access off Gloucester-Walcha Road. Mountainous, tall eucalypts, rainforests and bird life. Tel. (049) 87 3108.

Wollemi National Park—4000 sq km, 50 km north-west of Sydney. Car access to park boundary along Windsor-Singleton and Lithgow-Rylstone Roads. Massive area of unaltered forests, waterways, sandstone canyons, steep gorges, rugged bushwalking. Tel. (065) 43 2274, (047) 87 8877, (045) 75 1671.

Yuragir National Park—13 000 ha, 40 km south of Grafton. Car access off the Pacific Highway. Beaches, headlands, lagoons, rivers, plant and bird life, water sports and bushwalking. Tel. (066) 42 0613.

Queensland

Alton National Park—560 ha, 440 km west of Brisbane. Car access via Moonie Highway. Hardy woodlands, sandplain and heath, wildflowers. Tel. (076) 38 1897.

Archer Bend National Park—166 000 ha, 750 km north-west of Cairns. Access in four-wheel-drive only off Peninsula Developmental Road, permission required. Archer River floodplain, rainforests, swamps, lagoons, waterfowl and crocodiles. Tel. (070) 53 4533.

Auburn River National Park—400 ha, 385 km south of Rockhampton. Car access off Burnett Highway at Mundubbera, some creek flooding. Woodlands bisected by Auburn River, cascades and rockpools, walking trails and lookouts. Tel. (071) 22 2455.

Barron Gorge National Park—2800 ha, 30 km north-west of Cairns. Car access off Kennedy Highway at Kuranda. Railway winds through rainforest, superb views of gorge and falls. Tel. (070) 53 4533.

Bendidee National Park—930 ha, 360 km south-west of Brisbane. Car access off Cunningham Highway. Undeveloped scrublands, wildflowers, large bird population. Tel. (076) 84 5157.

Blackall Range National Park—181 ha, 22 km west of Nambour. Car access off Nambour-Kenilworth Road west of Mapleton. Waterfalls, cascades and remnant rainforests in heavily timbered mountain range. Tel. (071) 82 4189.

Black Mountain National Park—780 ha, 30 km south of Cooktown. Car access via Cairns-Cooktown Road skirting western boundary. Habitat for Queensland rock wallaby, rocky summits, some rainforest. Tel. (070) 53 4533.

Blue Lake National Park—500 ha, 60 km east of Brisbane on North Stradbroke Island. Car access via vehicle ferry from Redland Bay and Cleveland. Island coastline, freshwater lakes, eucalypt woodlands. Tel. (07) 202 7000.

Bowling Green Bay National Park—55 000 ha, 30 km south-east of Townsville. Car access off Bruce Highway, four-wheel-drive in wet conditions. Steep wilderness range, Alligator Creek, coastal mudflats and mangrove forests. Tel. (077) 74 1411.

Brampton Island National Park—464 ha, 40 km north-east of Mackay. Accessible by daily flights and launch from Mackay. Continental island, tourist resort, water sports, coral reefs and surrounding island parks. Tel. (079) 57 6892.

Burleigh Head National Park—24 ha, 15 km south of Southport. Car access off Pacific Highway. A steep headland at Tallebudgera Creek entrance, short nature trails, forest. Tel. (075) 35 3183.

Cania Gorge National Park—1000 ha, 250 km south of Rockhampton. Car access off Burnett Highway north-west of Monto. Gorge with narrow side gullies, rainforest, walking trails, lookouts and caves. Tel. (071) 22 2455.

Cape Hillsborough and Wedge Island National Park—800 ha, 45 km north-east of Mackay. Car access off Bruce Highway to Cape Hillsborough, Wedge Island can be reached on foot at low tide. Dunes, coastal scrub, rainforest, swimming. Tel. (079) 59 0222.

Cape Melville National Park—36 000 ha, 200 km north of Cooktown. Access by four-wheel-drive from Cooktown via Starcke Station and boat from Cooktown. Coastal scenery, swamps, heaths, Melville Range. Tel. (070) 53 4533.

Cape Palmerston National Park—7000 ha, 275 km north of Rockhampton, four-wheel-drive access from Bruce Highway at Ilbilbie. Coastline, sandy beaches, lagoons, range of hills. Tel. (079) 57 6292.

Castletower National Park—500 ha, 155 km south-east of Rockhampton. Car access off Bruce Highway at Iveragh siding (dirt). Steep granite outcrop, difficult access, bushwalking challenge. Tel. (079) 74 5238.

Cedar Bay National Park—5600 ha, 50 km south of Cooktown. Car access south from Cooktown along rough track, boat access. Mountain creeks, gorges, rainforest and bird life. Tel. (070) 53 4533.

Coalstoun Lakes National Park—26 ha, 360 km south of Cayndah. Foot access or four-wheel-drive from Isis Highway (3 km). Volcanic cone and craters, lakes and vine forest. Tel. (071) 22 2455.

Conondale National Park—2000 ha, 140 km north-west of Brisbane. Car access from Conondale-Kenilworth Road off Bruce Highway. Tall rainforest, mountainous slopes, undeveloped wilderness. Tel. (07) 20 2700.

Crows Nest Falls National Park—1000 ha, 50 km north of Toowoomba. Car access to north, east from Crows Nest (unsealed). South reached on foot off New England Highway. Ridges, heathland, rainforest, waterfalls and pools. Tel. (076) 21 3171.

Dagmar Range National Park—2500 ha, 100 km north-west of Cairns, Mossman-Daintree Road skirts eastern boundary, rainforest, ridges, streams. Tel. (070) 53 4533.

D'Aguilar Range National Parks—1328 ha, 40 km north-west of Brisbane. Four parks, mountainous rainforest, walking trails, views. Tel. (07) 289 0122.

Daintree National Park—56 000 ha, 80 km north of Cairns. Car access to Mossman Gorge, unsealed road to south-west of park. Wilderness park, upland tropical rainforest and woodland, steep scarps, waterfalls and Daintree River. Tel. (070) 53 4533.

Davies Creek National Park—480 ha, 60 km west of Cairns. Car access off Kennedy Highway between Mareeba and Kuranda. Davies Creek Falls, eucalypt forests, swimming holes, walking. Tel. (070) 53 4533.

Dipperu National Park—11 000 ha, 330 km north-west of Rockhampton. Car access from Peak Downs Highway. Numerous permanent lagoons, great variety of bird and animal life, cross-country walking. Tel. (079) 57 6292.

Dunk Island National Park—730 ha, 120 km south of Cairns. Flights from Cairns, launch from Clump Point. Busy resort island, beaches and bays, forest clad ridge. Tel. (070) 66 8601.

Ella Bay and Moresby Range National Parks—3400 ha and 240 ha, 100 km south of Cairns. Car access off Bruce Highway. Low mountain ranges lining rainforest east of Innisfail. Tel. (070) 53 4533.

Endeavour River National Parks—1840 ha, 5 km west of Cooktown. Car access from Cooktown. Mudflats, sand dunes, historical and botanical interest. Tel. (070) 53 4533.

Epping Forest National Park—2600 ha, 520 km north-west of Rockhampton. No public access. Protects sole colony of Queensland hairy-nosed wombat. Scientist entry only. Tel. (079) 27 6511.

Eubanangee Swamp National Park—1500 ha, 80 km south of Cairns. Car access off Bruce Highway at Miriwinni. Bird life refuge, swampland, lagoons, pools, rainforest. Tel. (070) 53 4533.

Eurimbula National Park—7300 ha, 220 km south-west of Rockhampton, four-wheel-drive access off Miriam Vale-Seventeen Seventy Road. Permit required, impassable after heavy rain. Beaches, creeks, swamps, subtropical and tropical vegetation. Tel. (079) 74 5238.

Fitzroy Caves National Park—100 ha, 22 km north of Rockhampton. Car access off Bruce Highway by 2 km dirt road. Limestone caves, rock formations, thousands of bats including the endangered ghost bat. Tel. (079) 27 6511.

Forty Mile Scrub National Park—4600 ha, 240 km south-west of Cairns. Car access via Kennedy Highway passing through park. Grassy eucalypt woodland, scrubland. Tel. (070) 953 7668.

Frankland Islands National Park—78 ha, 45 km south-east of Cairns. Boat access. Series of popular islands, white shady beaches, eucalypt forests, fringing coral reefs, resort at Bramston Beach. Tel. (070) 53 4533.

Freshwater Creek National Park—100 ha, 35 km north of Brisbane. Car access via Bruce Highway. Undeveloped wallum vegetation area, banksias, eucalypts, grass trees. Tel. (07) 202 7000.

Glasshouse Mountains National Parks—700 ha, 65 km north of Brisbane. Car access off Bruce Highway. Four parks, volcanic peaks, eucalypts, heath and rainforest, hiking. Tel. (07) 202 7000.

Gloucester and Middle Islands National Park—Chain of islands, 200 km south-east of Townsville. Gloucester Island (2460 ha) is the largest in the chain, fringing coral reefs, sheltered beaches, water sports, picnicking, camping. Tel. (079) 46 9430.

Graham Range National Park—2900 ha, 70 km south-east of Cairns, four-wheel-drive access only from Bramston Beach off Bruce Highway. Beach ridges, steep coastal slopes of Graham Range, Swampland, coastal vegetation. Tel. (070) 53 4533.

Great Sandy National Park—52 000 ha, 70 km north-east of Maryborough. Vehicular barge to near park boundary, four-wheel-drive only on island tracks. Northern section of Fraser Island. Sand dunes and beaches, lagoons, swamps, eucalypts. Tel. (071) 22 2455.

Green Island National Park—500 ha, 30 km north-east of Cairns. Access by daily boat services. Coral cay with surrounding marine park reef. Resort, underwater observatory, glass-bottom boats, beaches, water sports. Tel. (070) 53 4533.

Grey Peaks National Park—900 ha, 30 km south of Cairns. No vehicle access. Trackless rainforest, Grey Peaks ridge, bushwalking. Tel. (070) 53 4533.

Hasties Swamp National Park—50 ha, 85 km south-west of Cairns. Car access from Atherton off Herberton Road. Swamp, preservation of waterfowl. Tel. (070) 95 3768.

Heron Island National Park—17 ha, 70 km north-east of Gladstone. Air and launch access. Most popular of Capricorn islands, coral cay, tourist resort, marine research station, water sports. Tel. (079) 27 6511.

Hull River National Park—1100 ha, 155 km south of Cairns. Car access off Bruce Highway from Lower Tully (dirt). High foothills, creeks, swamps, mangrove flats, waterfowl. Tel. (070) 53 4533.

Iron Range National Park—35 000 ha, 800 km north of Cairns. Access in four-wheel-drive off Peninsula Developmental Road north of Coen, impassable after heavy rain. Bird and animal refuge in rainforest, eucalypt, mangrove wilderness. Tel. (070) 53 4533.

Isla Gorge National Park—7800 ha, 285 km south of Rockhampton. Foot access only off Leichhardt Highway south of Theodore. Steep cliffs, caves, arches, Isla Creek, rough tracks. Tel. (079) 27 6511.

Jourama National Park—1100 ha, 90 km north-west of Townsville. Car access off Bruce Highway (unsealed). Jourama Falls, Waterview Creek, river and rainforest vegetation.

Keppel Islands National Parks—60 km north-east of Rockhampton. Launch access from Yepoon. Eighteen continental islands, swimming, snorkelling, underwater observatory on Middle Island, resort on Great Keppel Island. Tel. (079) 27 6511.

Kroombit Tops National Park—450 ha, 170 km south of Rockhampton. Difficult foot access. Permit required. Calliope Range, hoop pine rainforest, waterfalls. Tel. (079) 27 6511.

Kurrimine National Park—900 ha, 60 km south-west of Cairns. Car access off Bruce Highway at Silkwood. Coastal fringe, dense rainforest, resort at Kurrimine Beach. Tel. (070) 53 4533.

Lake Barrine National Park—490 ha, 60 km south-west of Cairns. Car access via Gillies Highway passing through park. Volcanic crater lake, bird life, rainforest, swimming, lake cruises. Tel. (070) 95 3768.

Lake Eacham National Park—490 ha, 70 km south-west of Cairns. Car access off Gillies Highway. Rainforest, large volcanic crater lake, abundant flora and fauna, walking tracks and swimming. Tel. (070) 95 3768.

Lake Moogerah Peaks National Parks—610 ha, 110 km south-west of Brisbane. Car access via Boonah off Cunningham Highway. Three parks, Lake Moogerah, rocky cliffs, peaks, gorges, walking tracks and rock climbing. Tel. (07) 20 2700.

Littabella National Park—2400 ha, 40 km north-west of Bundaberg. No public access. Freshwater lagoon surrounded by flat heathlands, wildlife refuge. Tel. (071) 22 2455.

Lonesome National Park—3370 ha, 440 km south-west of Rockhampton. Car access south of Rolleston, gravel road crosses park. High cliffs loom over Dawson River, panoramic views, wildlife. Tel. (079) 82 2246.

Magnetic Island National Park—2500 ha, 10 km north of Townsville. Launch and vehicle ferry access from Townsville. Partly island park and partly urban Townsville. Swimming beaches, coral reefs, Mt Cook, walking, cycling. Tel. (077) 74 1411.

Main Range National Park—10 500 ha, 115 km south-west of Brisbane. Car access via Cunningham Highway crossing park. Steep ridges, mountain peaks, dense vegetation, abundant bird life and wildflowers. Tel. (076) 66 1133.

Maria Creek and Clump Mountain National Parks—970 ha, 120–140 km south of Cairns. Access on foot to Maria Creek. Car access to Clump Mountain on Bingil Bay-Mission Beach Road. Low palm swamps, creeks, views at Clump Mountain, walking tracks. Tel. (070) 53 4533.

Millstream Falls National Park—370 ha, 140 km south-west of Cairns. Car access off Kennedy Highway. Magnificent wide waterfall, banksia thickets, eucalypts, platypuses, swimming and walking. Tel. (070) 95 3768.

Mitchell and Alice Rivers National Park—37 000 ha, 670 km north-west of Cairns. Access in four-wheel-drive only via Kowanyama, impassable after heavy rains. Alluvial river plains, dense riverine vegetation and sparse woodlands. Tel. (070) 53 4533.

Mooloolah River National Park—670 ha, 95 km north of Brisbane. Car access from Tanawha off Bruce Highway, unsealed. Mooloolah River, swamps, sandflats, ridges to the north. Tel. (071) 82 4189.

Mt Aberdeen National Park—2900 ha, 40 km south-west of Bowen. Car access off Bowen Developmental Road. Steep slopes, huge granite boulders, red gum and hoop pine. Tel. (079) 46 9430.

Mt Barney National Park—11 500 ha, 130 km south-west of Brisbane. Car access off Mt Lindesay Highway, unsealed. Popular bushwalking from Logan River around Mt Barney, gorges, rapids, waterfalls. Tel. (075) 63 1579.

Mt Bauple National Park—540 ha, 35 km south of Maryborough. Car access from Bruce Highway, 2 km rough road. Mt Guyra, dense eucalypts, hoop pine rainforest. Tel. (071) 82 4189.

Mt Beatrice National Park—1140 ha, 55 km south of Proserpine. Foot access from Yalboroo. Steep slopes, seasonal streams, deep gullies, dense forest. Tel. (079) 57 6292.

Mt Blackwood National Park—1000 ha, 30 km north-west of Mackay. Foot access only from Seaforth Road at Kuttabul. Precipitous slopes, television tower on Mt Blackwood, surrounded by cane fields. Tel. (079) 57 6292.

Mt Colosseum National Park—840 ha, 180 km south-east of Rockhampton. Car access via Miriam Vale off Bruce Highway. Eastern face of Mt Colo is like the walls of a Roman arena. Panoramic views of coastline from summit. Tel. (079) 74 5238.

Mt Fox National Park—200 ha, 65 km south-west of Ingham. Car access west of Ingham, unsealed. Extinct volcanic crater of Mt Fox, 811 m around. Tel. (077) 74 1411.

Mt Hypipamee National Park—360 ha, 105 km south-west of Cairns. Car access off Kennedy Highway. Huge volcanic crater lake surrounded by towering rainforest, swimming at Dinner Falls.

Mt Jim Crow National Park—144 ha, 30 km north-east of Rockhampton. Car access via Rockhampton-Yepoon Road. Volcanic plug of Mt Jim Crow towering over eucalypt countryside, creeks and pools with waterlilies. Tel. (079) 27 6511.

Mt Major National Park—750 ha, 140 km south-west of Cairns. Foot access off Ravenshoe-Vine Creek Road. Massive bluff towering above steep slopes, vine jungle, waterfalls, timber preservation. Tel. (070) 95 3768.

Mt Mistake National Park—5500 ha, 120 km south-west of Brisbane. Access in four-wheel-drive only from Allora, north of Warwick. Steep slopes, logging tracks, plundered rainforest, bushwalking. Tel. (076) 66 1133.

Mt Tempest National Park—9000 ha, 50 km north-east of Brisbane on Moreton Island. Vehicle and passenger ferry and private boat access. Surf beaches, sand dunes, woodlands, four-wheel-drive tracks on island. Tel. (072) 48 2710.

Mt Walsh National Park—3000 ha, 350 km south of Rockhampton. Foot access through private property off Biggenden-Maryborough Road, permission required. Rock climbing peak, eucalypt forest, deep swimming pool. Tel. (071) 22 2455.

Natural Arch National Park—200 ha, 110 km south of Brisbane. Car access via Nerang-Murwillumbah Road. Natural arch rock formation, waterfall, swimming hole, rainforest, walking tracks. Tel. (075) 33 6156.

Noosa National Park—430 ha, 155 km north of Brisbane. Car access to Noosa Heads off Bruce Highway. Shoreline park between Noosa and Sunshine Beach. Walking tracks, lookouts, wildflowers, water sports. Tel. (071) 47 3243.

North Coast Rail National Parks—250 ha, 100 km north of Brisbane. Car access off Bruce Highway between Kulangoor and Landsborough. Four parks preserving bushland around Brisbane-Cairns railway. Tel. (071) 82 4189.

Northumberland Islands National Parks—150–240 km north of Rockhampton. Accessible by boat from Mackay, not all islands have safe anchorages. Chain of islands, some with sandy beaches, eucalypts, vine scrub, rainforest and large waterbird population. Tel. (079) 57 6292.

Orpheus Island National Park—1400 ha, 75 km north-west of Townsville. Boat access from Lucinda. Small tourist resort, coral reefs, sandy beaches, water sports. Tel. (070) 66 8601.

Palmerston National Park—2550 ha, 125 km south of Cairns. Car access via Palmerston Highway crossing park. High dense rainforest, deep gorges, rivers, waterfalls, abundant bird life. Tel. (070) 53 4533.

Peak Range National Park—1750 ha, 400 km north-west of Rockhampton. Car access from Peak Downs Highway, unsealed. Volcanic peaks, bushwalking, rock climbing. Tel. (079) 27 6511.

Porcupine Gorge National Park—2600 ha, 430 km west of Townsville. Car access along Kennedy Developmental Road, unsealed and impassable after heavy rains. Year-round watercourse, gorges, colourful wildlife. Tel. (077) 74 1411.

Possession Island National Park—500 ha, 980 km north of Cairns. Boat access from Bamaga and Thursday Island. Mangrove swamps, sandy beaches, obelisk commemorating visit by Captain Cook in 1770. Tel. (070) 53 4533.

Queen Mary Falls National Park—78 ha, 210 km south-west of Brisbane. Car access off Warwick-Legume Road. Spring Creek headwaters, cascades, lookouts, rainforest. Tel. (076) 66 1133.

Ravensbourne National Park—100 ha, 45 km north-east of Toowoomba. Car access off Hampton-Esk Road. Tall red cedar, rosewood forests, walking tracks, panoramic views. Tel. (076) 38 1897.

Reliance Creek National Park—14 ha, 10 km north-west of Mackay. Car access along Habana Road from Mackay. Small palm-vine forest area around Reliance Creek. Tel. (079) 57 6292.

Robinson Gorge—8900 ha, 445 km south-west of Rockhampton. Car access to north end of park, four-wheel-drive access off Leichhardt Highway. Aboriginal art, gorges, swimming, wildflowers and fauna. Tel. (079) 27 6511.

Rocky Creek National Park—70 ha, 190 km south-west of Brisbane. Foot access east of Warwick at Emu Vale. Undeveloped rainforest, gorges, waterfalls, scenic views. Tel. (076) 66 1133.

Rokeby-Croll Creek National Park—290 000 ha, 600 km north of Cairns, four-wheel-drive access along Peninsula Developmental Road, impassable after heavy rain. Long, winding rivers and floodplains, abundant bird life, crocodiles. Tel. (070) 53 4533.

Saint Helena Island National Park—75 ha, 8 km north-east of Brisbane. Access by organised tours from Manly wharf. Ruins of nineteenth-century prison, church, etc. Tel. (07) 202 7000.

Snake Range National Park—1210 ha, 400 km west of Rockhampton. No public access. Natural amphitheatre, waterways, eucalypts, caves, Aborginal art. Tel. (079) 82 2246.

Southwood National Park—7100 ha, 130 km south-west of Dalby, Darling Downs. Car access via Moonie Highway. Brigalow and acacia scrubland, wildflowers. Tel. (076) 38 1897.

Springbrook National Parks—2070 ha, 100 km south of Brisbane. Car access from Mudgeeraba off Pacific Highway. A chain of four parks, waterways rising in steep plateaux, rapids, cascades, pools, rainforest, views to coast. Tel. (075) 33 5147.

Staaten River National Park—467 000 ha, 450 km west of Cairns, four-wheel-drive access only off Burke Developmental Road, permission required, impassable after heavy rain. Vast open plains, huge termite mounds, rainforest, rivers. Tel. (070) 53 4533.

Starcke National Park—8000 ha, 80 km north of Cooktown, four-wheel-drive access only from Cooktown, impassable after heavy rains. Steep weathered ridges, heaths, scrub, rugged Cape York Peninsula country. Tel. (070) 53 4533.

Sundown National Park—6700 ha, 85 km south-west of Stanthorpe-Glenlyon Road in dry conditions. Deeply dissected granite range, gorges, meandering Severn River, bushwalking, swimming. Tel. (067) 37 5235.

Tamborine Mountain National Park—650 ha, 75 km south of Brisbane. Car access off Oxenford–Tamborine Village Road. Fragmented groups of parks, rainforest, waterfalls, swimming holes, picnicking, bushwalking. Tel. (075) 45 1175.

The Palms National Park—12 ha, 230 km north-west of Brisbane. Car access off New England Highway. Picabeen palms, tree ferns, nature trail. Tel. (076) 38 1897.

Topaz Road National Park—37 ha, 97 km south of Cairns. Car access via Lamins Hill, partly unsealed. Steep gullies, creeks, remnant rainforest. Tel. (070) 95 3768,

Tully Gorge National Park—500 ha, 160 km south of Cairns. Car access from Ravenshoe. Steep rainforested gorge, waterfalls in wet season, walking track and lookout. Tel. (070) 95 3768.

Wallaman Falls National Park—600 ha, 160 km north-west of Townsville. Car access west of Ingham, unsealed. Wilderness, impressive 278 m waterfall, rapids, gorges, rainforest, lookouts and swimming holes. Tel. (077) 74 1411.

West Hill National Park—680 ha, 250 km north of Rockhampton, four-wheel-drive access via Bruce Highway and Orkabie, boat access. Coastal scenery, rainforest, mangroves. Tel. (079) 57 6292.

Woodgate National Park—5500 ha, 330 km south-east of Rockhampton. Car access off Bruce Highway, four-wheel-drive for park travel. Rivers, beaches, prolific bird life, swimming and fishing. Tel. (071) 22 2455.

Yepoon National Park—107 ha, 50 km north-east of Rockhampton. Car access between Yepoon and Emu Park, daily buses. Wide beaches, headlands, views to Keppel Bay islands. Tel. (079) 27 6511.

Victoria

Alfred National Park—2300 ha, 470 km east of Melbourne. Car access via Princes Highway crossing park. East Gippsland rainforest, wattles, granite ridge. Tel. (051) 58 6351.

Bogong National Park—81 000 ha, 357 km north-east of Melbourne. Car access off Ovens Highway via Mt Beauty or Harrietville. Cross-country skiing, bushwalking, camping, spectacular scenery and summer wildflowers. Tel. (057) 57 2693.

Brisbane Ranges National Park—7517 ha, 105 km west of Melbourne. Car access along Ballan-Geelong Road off Western Freeway. Wildflowers, bird life, Anakie Gorge and Steiglitz Historic Park, bushwalking and camping. Tel. (052) 84 1230.

Bulga National Park—80 ha, 30 km south of Traralgon. Car access off Princes Highway at Traralgon. Tall eucalypt forests and fine tree ferns, bird life and easy walking tracks. Tel. (051) 96 6127.

Churchill National Park—193 ha, 32 km south-east of Melbourne. Car access off Stud Road, Rowville. Dandenong foothills woodland, native fauna, bellbirds, walking trails. Tel. (03) 700 4700.

Ferntree Gully National Park—466 ha, 34 km east of Melbourne. Car access off Burwood Highway. Trains from Melbourne. Excellent picnic facilities, sweeping views of city, eucalypts and tree ferns. Tel. (03) 758 1001.

Glenaladale National Park—183 ha, 60 km north-west of Bairnsdale, East Gippsland. Car access off Princes Highway via Fernbank. Den of Nargun, a rocky cave on the Mitchell Road, lush rainforest, walking tracks to lookouts. Tel. (051) 52 2454.

Lind National Park—1166 ha, 425 km west of Melbourne, East Gippsland. Car access off Princes Highway. Scenic drive, rainforest, spring wildflowers, walking trails. Tel. (051) 58 6351.

Mount Eccles National Park—400 ha, 45 km south of Hamilton. Car access off Hamilton-Port Fairy Road. Series of crater lakes, lava canals, caves and stony rises. Nature walks around crater rims. Tel. (055) 76 1014.

Mount Richmond National Park—1707 ha, 33 km west of Portland. Car access off Portland-Nelson Road. Extinct volcano covered by sand, profusely flowering heaths, birds, native fauna. Tel. (055) 23 3232.

Organ Pipes National Park—85 ha, 25 km north-west of Melbourne. Car access off Calder Highway at Sydenham. Towering basalt columns and other volcanic formations. Tel. (03) 390 1082.

Snowy River National Park—42 000 ha, 135 km north of Lakes Entrance. Car access from Princes Highway off Buchan-Jindabyne Road along Bongang Road. High peaks, deep river valleys, gorges, demanding canoeing and bushwalking. Tel. (0648) 80 277.

Tarra Valley National Park—140 ha, 35 km south of Traralgon. Car access off Princess Highway at Traralgon. Ferny rainforests, waterfalls, walking tracks. Tel. (051) 96 6127.

Tingaringy National Park—44 000 ha, 450 km north-east of Melbourne near New South Wales/Victoria border. Car access north of Buchan. Mountainous terrain, fast-flowing rivers, narrow gorges, snow skiing and summer hiking. Tel. (0648) 80 277.

Wonnangatta-Moroka National Park—107 000 ha, 140 km north of Sale. Car access off Heyfield-Mansfield Road. Mountain peaks, deep river valleys, ski-touring, bushwalking, camping. Tel. (051) 74 6166.

South Australia

Belair Recreation Park—800 ha, 13 km south of Adelaide. Sportsfields, golf course, adventure playground, tennis courts and a kiosk. Deep valleys, pink, blue and mauve gums with spring wildflowers seen on walks. Tel. (08) 278 5477.

Black Hill Conservation Park—1300 ha, 12 km north-east of Adelaide. Eucalypt forests and woodlands, rough ridges and steep gorges. Superb waterfalls after rain. Educational walks to view native plant park and nursery. Rock climbing excellent. Tel. (08) 336 3966.

Cape Gantheaume and Seal Bay Conservation Parks—700 ha on south-eastern coast of Kangaroo Island. Car access by ferry from Adelaide. Sea lions, fur seals and occasionally leopard seals as well as thousands of waterbirds. Picnic shelters, barbecues, toilets and water. Bushwalking permitted but swimming and fishing prohibited at Seal Bay. Tel. (0848) 22 540.

Cleland Conservation Park—9 km south-east of Adelaide. Timbered ridges and rocky gorges. Kangaroos and emus wander freely. Excellent walks. Tel. (08) 216 7777.

Coffin Bay—30 000 ha, 50 km west of Port Lincoln. Limestone cliffs, outcrops of smooth granite, dunefields and ridges. Flowering heathlands can be seen on walks. Tel. (086) 85 4047.

Deep Creek Conservation Park—4000 ha, southern tip of Yorke Peninsula. Car access from Delamere. Adelaide rosellas, yellow-tailed black cockatoos, inland forests of stringybark and messmate. Steep walking trails. Camping only with permit. No water in park. Tel. (085) 59 2263.

Gammon Ranges Conservation Park—100 000 ha, 265 km north of Hawker. Blue-grey mulga and casuarinas, sparse slopes and tree-lined creeks. Flocks of parrots and yellow-footed rock wallabies. Fee for camping. Carry drinking water. Tel. (086) 42 3800.

Innes Conservation Park—9000 ha, 145 km south of Androssen. Car access off Highway 1 at Port Wakefield. Home of the rare western whipbird and other birds. Mallee woodland, salt marshes and salt lakes. Strong winds often prevail. Fee for camping. Carry drinking water. Tel. (011), ask for Shenhouse Bay 12.

Kelly Hill Conservation Park—on the south-western coast of Kangaroo Island. Car access by ferry from Adelaide. Caves, occasional sinkholes, ancient sand dunes and lagoon. Camping only with permits. Excellent walks. Tel. (048) 37 231.

Naracoorte Caves Conservation Park—300 ha, 14 km south of Naracoorte. Car access by sealed road. Fossil caves, bent wing bats and river red gums. Camping only with permits. Tel. (087) 62 2340.

Nullarbor National Park—230 000 ha, 300 km west of Ceduna. Car access along Eyre Highway. Patchy mallee scrub and low ground cover. Grass and wildflowers after rain. Hairy-nosed wombats emerge at night. Spectacular views of the Great Australian Bight from side roads. Tel. (086) 76 1098.

Tantanoola Conservation Park—20 km south-east of Millicent. Series of caves, massive columns and spectacular helictites. Walking trail to park lookout. Tel. (087) 34 4153.

Northern Territory

Emily and Jessie Gaps National Park—700 ha, 12 km east of Alice Springs via Ross Highway (sealed road). Two gaps, sparse vegetation, wildflowers and Aboriginal paintings. Tel. (089) 50 8211.

Gurig National Park—220 000 ha, 200 km north-east of Darwin through Arnhem Land. Restricted four-wheel-drive access, usually impassable December–June. Abundant bird life, saltwater crocodiles, water buffalo and tropical eucalypt forest. Tel. (089) 22 0211.

Western Australia

Alexander Morrison National Park—8501 ha, 270 km north of Perth. Car access via Brand Highway (sealed) and Coorow Road (gravel). Sandplain heaths, wildflowers and bird life. Tel. (09) 52 5051.

Avon Valley National Park—4377 ha, 80 km north-east of Perth. Car access off Toodyay Road via Morangup Hill (gravel). Impressive views of Avon River, waterfall, jarrah and marri forests. Tel. (095) 26 2540.

Badgingarra National Park—13 121 ha, 205 km north of Perth. Spring-flowering heaths preserve the rare black kangaroo paw and other restricted species. Tel. (095) 45 7043.

Boorabin National Park—26 000 ha, 465 km north-west of Esperance near goldfields. Sandplain heath with dazzling spring wildflowers. Tel. (090) 386 8811.

Collier Range National Park—277 841 ha, 600 km south of Port Hedland. Dirt access road off Great Northern Highway, sparse spinifex grass growth, sandplains and dunes with eucalypts along creek beds. Tel. (091) 86 8291.

D'Entrecasteaux National Park—36 599 ha, 350 km south of Perth on south coast. Limited four-wheel-drive access off Nannup-Pemberton Road. Wetlands, forests and coastal dunes rich in plant species. Tel. (097) 76 1042.

Drovers Cave National Park—2681 ha, 250 km north of Perth. Four-wheel-drive access off Brand Highway and Jurien Road. Caves, prehistoric mammal bones, sandplain heath and woodland. Tel. (095) 45 7043.

Drysdale National Park—435 591 ha, 350 km north-west of Kunnanurra on north-west coast of Western Australia. No public access. Isolated wilderness between Ashton Range and Carson Escarpment. Tel. (091) 80 8291.

Eucla National Park—3342 ha, 930 km east of Esperance near Western/South Australia border. Car access via Eyre Highway. Four-wheel-drive through park land. Sand dunes, mallee scrub, impressive views of Great Australian Bight. Tel. (09) 386 8811.

Frank Hann National Park—49 877 ha, 230 km north-west of Esperance in goldfields district. Car access via Norseman-Lake King Road (gravel). Spring wildflowers, mallee scrub and sandplains. Tel. (098) 35 5043.

Goongarrie National Park—49 878 ha, 100 km north of Kalgoorlie. Car access via Menzies Road (gravel). Mulga dominated vegetation. Ephemeral wildflowers in remote outback region. Tel. (09) 386 8811.

Gooseberry Hill National Park—33 ha, 22 km east of Perth. Car access via Lascelles on western edge of Darlington Range escarpment. Sightseeing road known as Zig-zag. Prolific spring wildflowers. Tel. (09) 198 8244.

Greenmount National Park—56 ha, 25 km east of Perth. Car access off Great Eastern Highway via Padbury Road. Panoramic views of coastal plain, Swan River and Perth. Jarrah trees and wildflowers. Tel. (09) 298 8244.

Hassell National Park—269 ha, 50 km north-east of Albany. Car access via South Coast Highway between Albany and Bremier Bay. Wildflowers and prolific bird life. Tel. (098) 27 9230.

Kalamunda National Park—375 ha, 30 km east of Perth. Car access via Kalamunda to Spring Road and Schipp Road. Plant reserve, mainly undeveloped. Tel. (09) 298 8244.

Lesmurdie Falls National Park—56 ha, 25 km east of Perth. Car access along Lesmurdie Road, through Forestfield or Lesmurdie. Popular picnic spot, good views of coastal plain, waterfall. Tel. (09) 298 8244.

Moore River National Park—17 543 ha, 113 km north of Perth. Car access via Brand Highway. Outstanding wildflowers, undeveloped. Tel. (095) 61 1004.

Neerabup National Park—1078 ha, 35 km north of Perth. Car access via Wanneroo Road and Quinns Rocks Road. Limestones of old dune systems, jarrah, tuarts and wildflowers. Tel. (095) 61 1004.

Peak Charles National Park—39 959 ha, 185 km north of Esperance in goldfields region. Car access via Coolgardie-Esperance Highway from Kumarl (gravel tracks, impassable after rain). Spectacular granite outcrops, scenic panorama. Tel. (090) 76 8541.

Tasmania

Asbestos Range National Park—4281 ha, 60 km north-west of Launceston. Car access on West Tamar Highway and Badger Head Road. Sandy beaches, coastal heathlands, dry eucalpyt woodlands, flora and fauna. Tel. (004) 28 6277.

Franklin-Lower Gordon Wild Rivers National Park—195 200 ha, 200 km north-west of Hobart, 42 km south-east of Queenstown. Car access via Lyell Highway. Remote wilderness area, rainforest, river cruising, rafting, canoeing. Tel. (004) 71 7122.

Hartz Mountains National Park—6470 ha, 85 km south-west of Hobart. Car access from Geeveston off Huon Highway via Arve and Hartz Road, last section gravel sometimes closed by snow. Alpine moorlands, picturesque lakes, wildflowers, walking tracks. Tel. (002) 30 8033.

Hellyer Gorge State Reserve—569 ha, 100 km west of Devonport. Car access through park on Murchison Highway via Somerset. Steeply winding gorge, rainforest, rivers. Tel. (003) 41 5312.

Mt William National Park—13 806 ha, 160 km east of Launceston. Car access via Tasman Highway and Herrick-Gladstone Road. Beaches and bays, wildlife reserve for Forester kangaroo, seals, reptiles, birds. Tel. (003) 54 2122.

Rocky Cape National Park—3050 ha, 100 km west of Devonport. Car access via Bass Highway. Coastal scenery and heath wildflowers, bird life, Aboriginal relics. Tel. (003) 41 5312.

Strzelecki National Park—4215 ha, 165 km north-east of Launceston. Air access only from Launceston and Hobart. Granite peaks, coastal plains, sand dunes and beach. Tel. (003) 59 2148.

Tasman Peninsula State Reserve—138 ha, 80 km south-east of Hobart. Car access to Port Arthur, the Blowhole, Tasmans Arch. Impressive coastal scenery, walking tracks. Tel. (003) 50 2107.

Truchanas Nature Reserve—400 ha, 200 km west of Hobart. No vehicle access, river access by raft. Huon pine preservation, gentle slopes, Denison River. Tel. (003) 30 8033.

Walls of Jerusalem National Park—11 510 ha, 110 km west of Launceston. Car access on Bass Highway to Deloraine then unsealed access road. Spectacular sheer cliffs, lakes and tarns, pencil pines, fauna and subalpine flora. Tel. (003) 41 5312.

Index